S. Hrg. 103–735

ENERGY DEMAND IN CHINA AND OTHER RAPIDLY DEVELOPING COUNTRIES IN ASIA

HEARING

BEFORE THE

COMMITTEE ON ENERGY AND NATURAL RESOURCES UNITED STATES SENATE

ONE HUNDRED THIRD CONGRESS

SECOND SESSION

ON THE

ENERGY DEMAND IN CHINA AND OTHER RAPIDLY DEVELOPING COUNTRIES IN ASIA

MARCH 16, 1994

Printed for the use of the
Committee on Energy and Natural Resources

U.S. GOVERNMENT PRINTING OFFICE

82–926 CC WASHINGTON : 1994

For sale by the U.S. Government Printing Office
Superintendent of Documents, Congressional Sales Office, Washington, DC 20402
ISBN 0-16-045893-5

COMMITTEE ON ENERGY AND NATURAL RESOURCES

J. BENNETT JOHNSTON, Louisiana, *Chairman*

DALE BUMPERS, Arkansas
WENDELL H. FORD, Kentucky
BILL BRADLEY, New Jersey
JEFF BINGAMAN, New Mexico
DANIEL K. AKAKA, Hawaii
RICHARD C. SHELBY, Alabama
PAUL WELLSTONE, Minnesota
BEN NIGHTHORSE CAMPBELL, Colorado
HARLAN MATHEWS, Tennessee
BYRON L. DORGAN, North Dakota

MALCOLM WALLOP, Wyoming
MARK O. HATFIELD, Oregon
PETE V. DOMENICI, New Mexico
FRANK H. MURKOWSKI, Alaska
DON NICKLES, Oklahoma
LARRY E. CRAIG, Idaho
ROBERT F. BENNETT, Utah
ARLEN SPECTER, Pennsylvania
TRENT LOTT, Mississippi

BENJAMIN S. COOPER, *Staff Director*
D. MICHAEL HARVEY, *Chief Counsel*
G. ROBERT WALLACE, *Staff Director for the Minority*
GARY G. ELLSWORTH, *Chief Counsel for the Minority*

CONTENTS

STATEMENTS

APPENDIXES

Appendix I

Appendix II

ENERGY DEMAND IN CHINA AND OTHER RAPIDLY DEVELOPING COUNTRIES IN ASIA

WEDNESDAY, MARCH 16, 1994

U.S. SENATE,
COMMITTEE ON ENERGY AND NATURAL RESOURCES,
Washington, DC.

The committee met, pursuant to notice, at 9:41 a.m., in room SD–366, Dirksen Senate Office Building, Hon. J. Bennett Johnston, chairman, presiding.

OPENING STATEMENT OF HON. J. BENNETT JOHNSTON, U.S. SENATOR FROM LOUISIANA

The CHAIRMAN. The hearing will come to order. This is an oversight hearing on the impact of growth in energy demand in China and other rapidly developing countries in the Asia-Pacific region.

This past January, I led a seven member Senate delegation—all but one member of the delegation serves on this committee—to China, Vietnam, Thailand, and Indonesia. These countries all have rapidly growing economies with major energy infrastructure development needs. Energy growth of the magnitude this region is experiencing has significant implications for the environment as well as international fuel and capital markets.

Last year, the committee held a hearing focused on the energy needs of the People's Republic of China. Due to its sheer size, China far outweighs any other country in the world with respect to the potential impact its growth will have on international energy markets and global environmental values. China's current non-transportation fuel mix is almost entirely coal and biomass, both of which, given the state of technology and current practices, are responsible for serious environmental degradation and public health problems. Clearly, given the vast coal reserves in China, coal will remain a major source of fuel. One of the questions before us today is how China can use these fuels and others to meet its energy needs with the least environmental impact.

Vietnam, the most recently emerging economy in the region, has made a strong commitment to developing its oil and gas resources. Beyond the obvious investment considerations for U.S. industry, an open question is whether the commitment will be made to developing natural gas to fill its potential as an environmentally benign fossil fuel. This has not always been the case in the region, in China, for example, because of the high cost of developing the transmission and distribution infrastructure.

Indonesia, by contrast, with population and economic growth continuing a pace, has watched its oil output steadily decline in recent years. As a result, Indonesia is predicted to be a net oil importer by the year 2000. To compensate, liquid natural gas, which has been a major export fuel, is being used more extensively domestically, and coal output has gone from less than 300,000 tons 10 years ago to 26 million tons in 1993. The state coal mining company plans to expand production to 60 million tons by the year 2000. While 15 to 20 million tons will supply export markets, the balance will be consumed domestically.

I think the nexus between the energy growth in the Asia-Pacific region, and the commitments the United States has made to address global environmental problems is clear. The administration's proposed fiscal year 1995 budget contains numerous programs to meet commitments under the Framework Convention on Climate Change, and we can make a comprehensive, good faith effort to reduce domestic emissions of greenhouse gases, but if China alone continues to rely on conventional coal technologies to meet its rapidly expanding electric generating base, our efforts will have been in vain.

To explore these issues we have a distinguished panel of witnesses who will present testimony on current administration policies and programs, an assessment of the status quo, and the private sector experiences working in the region.

And we will lead off with our good friend, Bill White.

[The prepared statement of Senator Wallop follows:]

PREPARED STATEMENT OF HON. MALCOLM WALLOP, U.S. SENATOR FROM WYOMING

Good Morning! May express my appreciation for this opportunity to examine trends in economic and energy development among Pacific Rim countries. A year ago we examined similar trends in China.

Above all else, developing countries require economic growth fueled by adequate energy and financial resources. Today, more than 50 percent of the human race lives without commercial supplies of energy; their per capita energy consumption is only 5 percent of that of North America.

While the Pacific region contains 34 percent of the world's population, it consumes 18 percent of world commercial, primary energy. Yet It contains 24 percent of proven coal reserves, 5 percent of proven natural gas reserves, and 4 percent of proven oil reserves. These reserves are unevenly distributed. Consequently, many countries will have to turn to energy imports for their economic development.

The message is clear, energy is a provider of basic needs and services. Either their consumption of energy is going to increase prodigiously or millions of people are going to be without even the most basic necessities for a reasonable life. The global challenge will be to provide the necessary energy services to sustain economic growth in an environmentally sound manner. The local priority, however, is to relieve poverty perhaps the greatest polluter of them all.

Regardless of the justification, what is clear is that major export opportunities exist for American businesses in such areas as environmental technologies and electric power generation and transmission. Each $1 billion in export sales not only adds twice that amount to our real GNP, but it creates 20,000 Jobs. Consequently, I look forward to hearing the opinions of today's witnesses on markets for energy goods and services in the Pacific Rim.

For example, the economic future of China depends on finding the means to use its abundant coal resources in a manner that is responsive to regional and global environmental concerns and consistent with sustainable economic development. Nowhere else in the world is the challenge greater, and nowhere else in the world is the need for clean coal technologies greater. Besides various issues surrounding access to these technologies, the overriding issue is how may of the developing countries can obtain the necessary financing.

My judgment is that the technological and competitive advantages that U.S. manufacturers now enjoy are being eroded by huge regulatory burdens and confiscatory tax rates in our own country. Our government could reduce capital costs through saner environmental regulation and AMT relief and increase competitiveness of exported goods by making some taxes rebatable at the border.

Moreover, major opportunities for the enhancement of our international competitiveness are being lost because of an unwillingness by various Federal agencies to work together; for example, through the Department of Commerce's Trade Promotion Coordinating Committee.

During the 1970's, the Congress directed DOE to undertake trade promotion initiatives through the Committee on Renewable Energy Commerce and Trade (CORECT). CORECT never assumed the pro-active role that was envisioned, however. Nevertheless, the Energy Policy Act of 1992 expanded DOE's mission to include support for the export of energy efficiency, clean coal and environmental technologies, as well as renewable energy. However, once again implementation of these statutory mandates is being frustrated by interagency turf battles that must cease if these programs are to be effective.

I look forward to hearing from representatives of the Administration and American business on what the Federal government can do to enhance the competitiveness of American businesses selling abroad.

Thank you, Mr. Chairman.

STATEMENT OF WILLIAM H. WHITE, DEPUTY SECRETARY, DEPARTMENT OF ENERGY

Mr. WHITE. Thank you, Mr. Chairman. We have a lengthy statement which will be submitted for the record.

You have been a leader in letting this town and this country know that we live in a world economy and we live in a world environment, and we must make our policies and reorder our priorities on that basis.

There is no place, as the members of this committee know, that is experiencing faster growth than China, and the Pacific Rim generally.

I thought I would share, Senator Johnston, a few salient points, give a couple of examples of how we're redirecting our efforts to take into account the mandates of EPACT and the guidance that you and others have given, and our awareness of this growing market.

[Chart.]

Mr. WHITE. I will turn first to energy. Maybe I could direct your attention to a couple of charts that at least interest me. On the left is one concerning the addition of electrical capacity in China that demonstrates some of the figures that you were talking about—a total of over 140 gigawatts of planned electrical generating capacity. Those turbines could be made in the United States. That architectural engineering work can be done in the United States. We have the state-of-the-art technology.

And when we look, for example, at fossil, and I tried to find exactly who is going to build a new coal plant here versus in China, we come up a little short. You see that chart over on my right.

This is a trend that we are all looking at very carefully. We have this unprecedented growth in the trade deficit with China to over $23 billion last year—the trade deficit. And when we look at the rest of the world, the rest of the world runs a substantial trade surplus with China. You would actually have to add the American number of $23 billion to that surplus figure of $12 billion to see that the rest of the world is doing very well in trade with China, and that the United States has a way to go.

The CHAIRMAN. Where is that coming from?

Mr. WHITE. A lot of it is capital goods and machine goods, about 34 percent of it, I believe. And as I asked my colleagues here, I asked that same question, and they have gotten that exact figure for me. So, instead of fumbling for it broken down by category I'll ask him to help me out and find it while I—Eric is showing you the other chart.

[Chart.]

Mr. WHITE. Mechanical electric is 34 percent, and basic metals and tools 17 percent.

The CHAIRMAN. Things that they could import from the United States.

Mr. WHITE. Things they could import from the United States.

This is another depiction of the potential. Eric, why do we not go to the next chart here in electrical power generation alone.

[Chart.]

Mr. WHITE. Then we see the significance of China, the growth in carbon emissions. Of course all these numbers, when you carry it out that far, they get speculative. The point being, though, that if we only address the problem at home without talking about it in China, then we will be short-sighted.

Now, what are we doing about this? We are beefing up our activities, as you are aware of, in a whole variety of areas because of their dependence, as you put it, on coal and biomass. We have been emphasizing the utilization of our clean coal technologies in China, and we have submitted a request, as part of the fiscal year 1995 budget request, which also covers getting some authority for fiscal year 1996, for funds which have already been authorized and appropriated in the clean coal budget, allowing those funds to be directed towards deployment of technologies for the export market.

That is what the people in the coal industry say is the most cost-effective way to expend some of these monies, given the relative additions of coal fired powerplants in the United States versus China. We are talking about clean coal technologies here.

On the nuclear front, our restrictions on the balance of plant that could be sold to China, such as turbines, has been a significant impediment to market penetration in China, and we have worked diligently within the Department of Energy to get those balance-of-plant restrictions eliminated. I believe last week the announcement was made in the Commerce Register that the balance of plant restrictions have been eliminated.

We are beefing up our program on energy efficiency and renewables as well, requesting for the CORECT program an additional $5 million in fiscal year 1995. Much of this is in the area of technical assistance, just letting people know the technologies that are available.

I would say, and I guess I am bound to say, that the effectiveness of some of these expenditures in turn will depend upon our ultimate position concerning the restrictions that have been imposed on OPIC and AID. And as you are aware, it is likely that those issues will be considered in the broader context of Most Favored Nation treatment.

The CHAIRMAN. Would you tell me how those are going to be considered as part of MFN? I mean, suppose MFN is extended.

Does that mean OPIC, AID and the other aid programs will be automatically lifted? Is there linkage between those?

Mr. WHITE. I cannot tell you what the direct linkage would be. I can tell you that action on those items and lifting the restrictions which you and many other people in this country believe would be a wise thing to do has been delayed so that they can be considered as part of the overall study that Winston Lord is doing.

The CHAIRMAN. What is the state of Winston Lord's study with respect to all of those, what I would call trade restrictions or failure to extend the trade incentives that are normally extended such as OPIC to other countries? Is he actively pursuing that study, or do you assume he is doing it, or do we know?

Mr. WHITE. He is actively pursuing the study. DOE is participating. Because of developments in China, recent development say within the last 2 weeks and Secretary Christopher's visit, I am not going to say anything which implies what our position would be on that, but that is actively being pursued.

The CHAIRMAN. Well, I understand well what is going on with Secretary Christopher and MFN, but if MFN, as I hope is the case, will be extended then the question is what happens to all of these other restrictions that account to a large extent for the fact that we have a huge trade deficit with China and the rest of the world has a trade surplus, and part of the reason we do is because we will not let our own companies export into China and get the kind of support that you get everywhere else in the world? I mean, is anybody home as far that problem is concerned?

Mr. WHITE. Yes. And I am advised that not only is Assistant Secretary Lord's group actively studying that, but that there is a proposal that is at the NSC on that.

The CHAIRMAN. Well, I would hope you all would weigh in on that at DOE, because it is just an incredible policy. I mean, I understand MFN and I understand the basis for that. I do not understand, if you do extend MFN, the basis for not granting OPIC and the other provisions.

Mr. WHITE. I cannot recall when we started our colloquy if I had ended my statement, but now is as good a time to end it if I have not.

I think we recognize, Mr. Chairman, what you said, and we understand what the interests are. This is one of the great export markets, the great export markets of the world. And we should do what we can as public servants to assist our citizens.

[The prepared statement of Mr. White follows:]

PREPARED STATEMENT OF WILLIAM H. WHITE, DEPUTY SECRETARY, U.S. DEPARTMENT OF ENERGY

Good morning, ladies and gentlemen. I am pleased to appear before the committee today to discuss the domestic and international implications of energy demand growth in China and the developing countries of the Pacific Rim. To say that demand for energy in China and the other countries of the region is growing rapidly is a major understatement. The entire region is growing rapidly—economically, industrially and demographically. Driven by this rapid economic expansion and the large and growing population, energy demand is increasing more rapidly in Asia Pacific than in any other region of the world.

The committee has asked me to focus my comments on the development of energy infrastructure. In the interest of being concise, most of my testimony will focus on China's energy infrastructure. But the conditions I will describe are generally present throughout the region. These conditions are forcing energy officials to seek a broad range of technical solutions that are economic, efficient and environmentally beneficial that can provide the infrastructure to cope with the economic expansion.

China's 1990-2000 plan called for economic growth of six percent per annum. Double-digit growth rates in 1992 and 1993 have caused China to increase its estimate to 8-9 percent per annum, taking into account strenuous current efforts by the central government to avoid overheating the economy. Total energy requirements will increase 4-5 percent per annum, while electric power growth is predicted to be 8-10 percent per annum.

China ranks in the low to middle of developing countries in terms of energy use per capita, but the sheer size of its economy and population makes China the world's third largest overall energy producer and consumer, trailing only the United States and the former Soviet Union.

U.S. GOALS FOR THE ASIA ENERGY MARKET

There are many interrelated objectives behind U.S. Government activities to promote U.S. involvement in Asia's growing energy market:

Increase U.S. exports and U.S. jobs;

Influence policies that have the potential to open these markets by eliminating trade and investment barriers;

Encourage privatization and economic reform;

Promote competitive trade and foreign investment;

Satisfy the demand for energy through improved energy efficiency and emphasize the use of environmentally compatible sources of energy;

Assist China in following through on their commitment to implement the framework convention on climate change;

Eliminate trade and investment barriers.

With proactive and innovative policies, the United States can achieve both its economic and environmental goals with moderate investments. Asia's growing energy market provides an outstanding opportunity for the United States to promote and export new innovative energy technologies while establishing sustainable energy and economic development in the developing countries.

U.S. AND CHINA POLICIES ON TRADE INVESTMENT

The United States encourages trade and investment throughout the world as a matter of policy. For most of the developing countries in southeast Asia, we do not place any restrictions on the private-sector activities. China is the exception. Over the last decade, for reasons of human rights and nuclear nonproliferation, restrictions and sanctions that affect the energy industry have been imposed or remain under consideration. The most important of these provisions are:

The agency for international development is prohibited from operating in China;

The trade and development agency and the overseas private development corporation are prohibited from obligating foreign assistance funds to China without presidential waiver;

The Export-Import Bank is restricted by statute to a $50 million loan ceiling without a presidential determination;

Recent policy restricting the use of tied-aid credits;

Until last week, a policy restriction on the export of balance-of-plant equipment for Chinese nuclear power plants;

A condition for presidential certification during the congressional approval of the U.S.-China agreement for nuclear cooperation which prohibits the sale of equipment for a nuclear power plant until the certification can be made;

Export controls on strategic commodities and technologies, such as high speed computers, numerical control machine tools, nuclear materials, and other dual use commodities; and

The requirement for an annual determination of most favored nation tariff status, based upon progress in human rights.

Our message for today, however, is that there is still a growing market and great opportunities for U.S. business exports into China despite these restrictions. We need to take advantage of the situations offered to us today.

We also need to work hard, within the administration and with China, to eliminate as many of these restrictions as soon as we can. The China policy group, led by Assistant Secretary of State Winston Lord, is charged with the responsibility to provide fresh options and policy. The Department of Energy is a active player in that group.

We should also look at the recent progress we have made. Last week, the Department of Commerce lifted the restrictions on the turbines, generators, and other equipment that make up the balance-of-plant for nuclear plants. Commerce has also revised its export restrictions to allow more computer technology trade.

The initial joint committee on commerce and trade between the U.S. and China will be held in Washington on April 13-14, 1994. A subgroup on energy technologies is to be formed to discuss trade and investment matters and provide recommendations. We can use this opportunity to inform them about U.S. technologies and encourage their utilization.

Our DOE-led export mission to China last June has already led to one contract to a firm which was on that mission. This contract involves a technical and economic feasibility study on a proposed 5,000-ton-per-day "clean coal refinery." Another firm on that mission has signed a letter of intent for a technical and economic feasibility study on a proposed integrated coal cleaning, coal slurry pipeline which will produce a minimum of 5,000,000-ton-per-year of clean coal products. Other contract negotiations are ongoing as a result of that mission.

The Export-Import Bank is working toward offering project financing in China.

DOE has committed to supporting a high efficiency and very clean integrated coal gasification combined cycle showcase project in China pending congressional funding appropriations. The Chinese identified this technology as having great potential in their country.

There is a rapidly growing trend throughout Asia, and the world, toward privatization of energy organizations. The Chinese government encourages foreign investment in the energy sector on the basis of mutual benefit and profit. Foreign investment in all energy sectors, including renewable energy sources and capture and use of coalbed methane, is now urgently needed for China and other Asian countries to cope with enormous power demand increases projected over the next decade.

The Chinese government has recently moved toward three important policy positions with respect to foreign investment in power plants. First, while the Chinese government understands that foreign developed power projects would be more costly than conventional Chinese power plants and that rate tariffs would have to reflect that fact, electricity prices will not be allowed to rise to a level above the consumer's capability to pay. Second, the Chinese partner in an investment will take the majority share in any project because, in the government's view, the power industry is fundamental to the infrastructure. Third, the Chinese do not believe a risk factor should be included as a cost in the electric power business. The Chinese believe that American companies often include inflated risk factors in their calculations of internal rate of return for project development. DOE, DOC and other U.S. officials are discussing these issues with Chinese officials in an effort to persuade them to consider the impact these policies might have on American investors now negotiating contracts in China.

A CHANGING APPROACH

To facilitate more rapid energy development, Asian governments are expected to further separate regulatory and enterprise functions. The Chinese have already undertaken a decentralization of control in energy with the creation of smaller, specialized ministries, i.e., the ministry of electric power, the ministry of coal, the China National Offshore Oil Corporation and the China National Petroleum Corporation. The results of this decentralization have been to empower the regions to assume greater autonomy in energy development decisions, especially in electric power development, and to allow for the development of private power projects in cooperation with foreign firms.

Because power is clearly a bottleneck in Asia's economic development, foreign participation is increasingly welcomed. In order to attract foreign capital, technology and management expertise, foreign investors are now allowed to own an equity share and operate power plants in the Asia-Pacific. Additionally, the Chinese have opened up new oil fields previously closed to foreign participation, including part of the Tarim Basin. Unlike many developing countries, China is demonstrating the ability to obtain fully subscribed financing from the private capital markets for power and energy projects.

To reduce its subsidy burden—which reached $2 billion last year—and spur efficiency gains, China fully liberalized coal prices in January 1994 and petroleum product prices will be deregulated in May of 1994. Market pricing initiatives are occurring in other Asian countries as well. This movement toward market prices is an additional factor that will promote western investment in Asia's energy development.

Let me turn now to a review of individual energy sector development.

COAL DEVELOPMENT

Coal constitutes nearly 75 percent of China's total energy consumption—the highest share among all major coal consuming countries. This situation is one that strongly differentiates China from the other developing Asian nations. The structure of coal use is also quite different from industrialized countries where the electricity sector is the dominant coal consumer. For instance, 85 percent of coal use in the United States is in electricity production compared to 29 percent in China, where over half (53 percent) of coal is used by smaller industrial users and 18 percent by households. Although the percentage of coal in China's energy mix is expected to decline from 74 percent in 1990 to 60 percent in 2010, the rapid growth forecast for energy consumption means that annual coal use will nearly double during that period.

In fact, China has announced plans to increase production of coal to 1.5 billion tons by 2000, up from 1.15 billion tons in 1993. Plans to renovate existing coal mines and develop the Shenfudongsheng Mine—which may contain as much as one-third of China's known coal reserves—will provide a market for U.S. coal mining, coal cleaning and preparation and safety equipment. United Nations' trade data indicate that the United States holds more than 20 percent of the Chinese market in these sectors. A Chinese perception that U.S. mining equipment requires much less maintenance than competing Japanese and European products makes it likely that U.S. firms will retain that share through the year 2000 in a market that could be worth more than $3 billion.

The extensive burning of unprocessed, uncleaned coal by outdated and inefficient technologies has caused serious environmental problems. In 1990, China accounted for over 10 percent of the world's carbon emissions, although its economy accounted for less than 2 percent of the world's gross domestic product.

OIL AND GAS DEVELOPMENT

China became a net oil importer at the end of 1993. Official statistics indicate that onshore production actually fell over the last two years, with only production from new fields in the South China Sea preventing an overall production decline. Total output has hovered around 3 million barrels per day since 1988. China is struggling to prevent oil production from slipping further at its aging major onshore oil fields (Daqing, Shengli, and Liaohe), which produce roughly 80 percent of total output, with enhanced oil recovery techniques, including chemical injection. China's search for substantial new fields has been largely unsuccessful.

Exploration has focused on remote regions of western China, but despite official claims of reserves in that area totaling 350 billion barrels (Saudi Arabia holds some 260 billion barrels), the region still lacks sizeable proven reserves. Even if large reserves are found, significant logistical barriers, including the lack of a pipeline to transmit oil to processing centers in eastern China, are likely to delay dramatic production increases before 2000.

U.S. firms' technical expertise is likely to allow them to continue to dominate China's market for imported oil and gas exploration and extraction technologies, particularly seismic survey, heavy drilling equipment, and enhanced oil recovery technologies. The United States holds 65 percent of that market, which may be worth between $5 and $8 billion over the next seven years.

ELECTRIC POWER DEVELOPMENT

The Department of Energy led a mission of 24 U.S. electric power equipment and service companies to China in June of last year, so I can speak with considerable confidence on this particular energy sector. In 1992, installed capacity was more than 160 gigawatts, with 120 gigawatts of thermal power and 40 gigawatts of hydropower. China is planning to add 90 to 105 gigawatts over the next seven years at a cost of some $100 billion, 75 percent of which will be provided internally.

In addition to a very ambitious program to expand China's thermal plant capacity and efforts to develop additional nuclear and hydroelectric power plants, the Chinese plan to promote the development of biogas, tidal energy, geothermal, wind power, and solar energy to satisfy needs for both power generation and heat supply.

Mine mouth generation and siting at transportation junctions are approaches the Chinese have told us will be utilized, along with higher-capacity transmission lines to help alleviate the problem of moving huge amounts of coal from producing provinces to energy consuming regions. Rural electrification will be stressed to service the 120 million people in the countryside who do not have access to electricity (which also provides a huge potential for renewable energy, discussed later).

The technology of considerable interest to the Chinese is coal gasification, especially integrated coal gasification combined cycle technology, because of its efficiency (45% vs. 25% for existing facilities), the environmental advantages and the fact that this technology will allow the continued use of China's substantial coal reserves while minimizing environmental problems. The department has requested funding from Congress in the FY 1995 budget to support a demonstration project in China deploying this technology. The project would be co-funded by U.S. industry and the Chinese. The funding the department will provide will reduce the upfront risks associated with introducing a new, advanced technology to a new market.

This "showcase" project is one part of a proposed three part DOE program approach to the Chinese coal-fired power generation market. The second part of the program consists of education and training for both U.S. and Chinese participants to improve understanding of Chinese market needs and U.S. technology capabilities. The third part is to co-fund with U.S. industry, project identification and front-end development activities to leverage U.S. risk capital for definition of more business opportunities in China.

There are 13 power networks in China. Five of these are large interprovincial networks. We expect the transmission and distribution segment of the electric power market to increase dramatically over the next several years. The need to reduce the demands placed on the transportation system caused by shipments of coal by developing mine-mouth power plants will increase demand for sophisticated high voltage transmission lines. DOE and DOC will bring a trade mission of electricity transmission officials to the United States in April of this year.

Nuclear power is in the early stages of development in China, as it is in most of the developing countries of the Asia-Pacific region. Japan, Korea, and Taiwan are no-table exceptions. China recently inaugurated a French-supplied nuclear plant at Daya Bay. It has also developed a domestically designed nuclear power technology. The first plant of this type entered commercial operation in 1992 at Quinshan. China also is talking to Russia about potential cooperation in the building of nuclear power stations. As concerns about the environmental impact increase, China's stated plans are to build new nuclear powered plants as rapidly as possible.

If China begins construction on 14 new nuclear power plants in the next ten years as anticipated, then full participation of U.S. firms could bring $5 billion to the United States. This represents 100,000 jobs over a ten year period. Up to now, most of this market has been closed to U.S. participation due to the inability to bring the agreement for nuclear cooperation into force. However, the administration last week removed unilateral controls for the export of balance-of-plant equipment, such as turbines and generators. Restrictions continue to bar exports of nuclear island equipment.

ENVIRONMENTAL ISSUES AND APPROPRIATE TECHNOLOGY

The government of China, at the highest levels, is taking environmental protection very seriously. The Chinese government has recently provided the U.S. government with a draft plan "Agenda 21," a plan for sustainable development. This plan proposed a phased-in approach for implementation of energy efficient technologies and advanced power generation technologies. The Chinese are looking to the U.S. for technical assistance to implement these environmental plans.

Energy efficiency, conservation and demand management are good technologies for China, and there is interest in the techniques that have been developed abroad. Conservation and environmental protection will be increasingly emphasized, but because of the large capital needs of the country, the Chinese will generally avoid installing expensive, high technology hardware, making country-appropriate technology important.

U.S. firms in the China energy efficiency market will be supported by the Beijing Energy Efficiency Center, officially opened in December 1993. The center intends to develop energy-efficiency policies and joint-ventures essential for China to make a transition to a more sustainable energy development pattern. The center was established by a consortia of U.S. research laboratories and the energy research institute of the state planning commission of China.

The development of a similar center focused on renewable energy technologies is also being planned by the Department of Energy and the Environmental Protection Agency. This center will act as a catalyst to commercialize renewable energy technologies. A formal market assessment of renewable energy technologies is being prepared to assist the center in its market development efforts. Its primary emphasis will be to assist U.S. renewable technology manufacturers enter the Chinese market.

To maintain economic growth in an atmosphere of social stability, the Chinese government has taken positive steps to improve living conditions and promote economic growth for peasants in the countryside. One of these steps is rural electrification. China's goals for rural electrification, which are similar to other Asian countries, are:

To enlarge and extend the power supply from the power grid to their vast rural areas;

To accelerate the development of small and medium-sized hydropower stations;

To go all out for the development of renewable energy, e.g., to develop wind power generation in those areas with good wind power potential; and,

To build small-sized coal-fired power stations in those remote areas with coal deposits but without power supply from the power grid.

The Department of Energy is currently co-funding the solar electric light fund (SELF) in its efforts toward rural electrification in western China. The program has established a revolving credit fund in the Gansu Province of China and has installed about 120 residential photovoltaic lighting systems to date, with plans to expand to 1000 systems.

Over the past three years, renewable energy has established a beachhead in Asian markets and among key institutions like the World Bank, but the industry must deal with the proven track record of fossil fuel technologies and the urgent need for large increases in power in the region.

An innovative Indonesian renewable energy-based hybrid power project offers a replicable model that could be applied to meet the growing energy needs of other Asian countries. This model goes well beyond the traditional solar home lighting systems. This hybrid system also provides 24-hour power to support light industrial, commercial and manufacturing development in rural areas for economic development. Each system is remotely monitored through telecommunications links between the systems and regional operation centers. These regional centers are linked by low-earth-orbit satellite links to the state utility and to the U.S. headquarters in Rockville, Maryland. This $20 million project was designed by a U.S. company in collaboration with the government of Indonesia.

The Indonesia model of providing renewable utility-grade AC power to community microgrids can address some of Asia's immediate needs for power without the environmental impacts associated with expanded rural electrification.

IMPLICATIONS FOR THE UNITED STATES

U.S. energy (electric power, oil, gas and environmental) firms are highly competitive in most of the technologies that I have will be utilized by the Chinese in their energy development program. The other Asia-Pacific countries have substantially the same energy interests as the Chinese. They will use a mix of technologies based on the availability and cost of fuel. We know from our contacts with energy officials in the region that there is a high degree of interest in learning more about U.S. energy technologies. The Department of Energy is supporting U.S. firms in the region with a program of activities to inform Asian energy officials about the benefits of U.S. based technologies and to promote the utilization of these technologies.

The U.S. has taken an aggressive leadership role in the Asia-Pacific Economic Cooperation (APEC) energy working group to promote linkages between regional economic growth and sustainable energy systems and to develop regional policy priorities and program options. The Department of Energy chairs the technical groups on the energy and environment theme (clean coal technologies) and on the energy efficiency and conservation themes. These groups have conducted seminars and study tours to facilitate technology adaptation, training, and information dissemination to support U.S. industry export efforts. The APEC energy group has already published several documents on APEC energy statistics, efficiency and renewable programs in the region, best efficiency practices compendium, and analysis of clean coal technology prospects.

The energy efficiency and conservation theme has advanced sufficiently that we are requesting $940,000 in the FY 95 budget. This contribution to APEC will be matched by contributions from the 12 other member countries. These funds will be used to support various cooperative efforts in the Pacific region. A workshop on congeneration or energy management will be held in Beijing in early 1995. The inter-utility DSM liaison group, which includes China will continue to explore issues common to all participants. APEC's renewable energy technology cooperation working group, which includes China, has identified three project proposals for FY 1995: (1) a workshop on sustainable development and renewable energy; (2) a renewable energy resource assessment for the region; and (3) an APEC sponsored renewable energy tour of the U.S.

The Asia Pacific region is in a period of long-term and rapid economic development. Rapid growth and demand for increased standards of living will drive the market for energy for the foreseeable future. What this means for the United States is an opportunity to export the outstanding technologies we have worked so hard to develop to solve our own energy challenges. U.S. energy firms are highly competitive and the products and services they provide are on the cutting edge of energy technology. They respond to today's needs for efficiency and environmental sensitivity. Some of these technologies were developed with taxpayer dollars in programs specifically designed to develop solutions for U.S. energy objectives. Deploying these technologies internationally will give the taxpayers something else for their money—high paying manufacturing and service jobs.

The Trade Promotion Coordinating Committee provides the framework and the export strategy for aggressive support by this administration. In addition, the Department of Energy has been working closely with the Departments of Commerce and State and the EPA in the formulation and implementation of our programs with China. The U.S. Embassy Beijing, including Ambassador Roy and the science and technology, economic and foreign and commercial service sections have been and continue to be instrumental in providing up to date information on the opportunities and considerations necessary to develop sound programs with our Chinese counterparts. It is through this team approach that we can be most effective in supporting U.S. industry as they attempt to penetrate this very large energy market. Other countries are highly competitive, have better geographic location, and provide strong government support for trade and investment in this region. We must use our strengths to the utmost to support our exporters.

China's Merchandise Trade Surplus with United States Increases

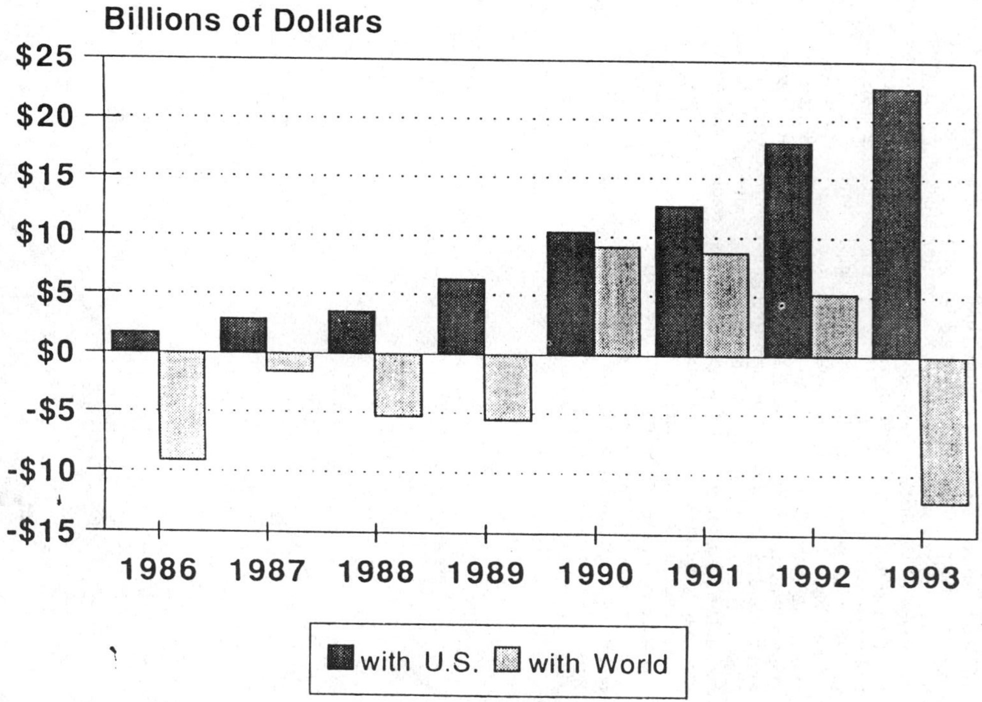

China's Electricity Capacity Grows Faster Than U.S. from 1990-2000 (Gigawatts)

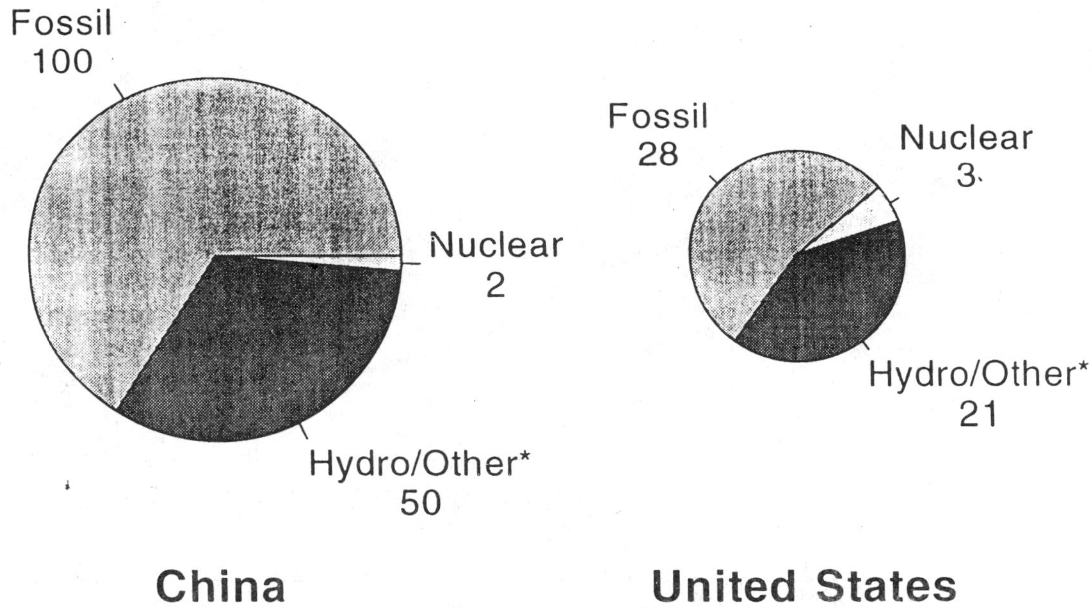

*Includes cogeneration

China's Energy Intensity Falls but Remains High

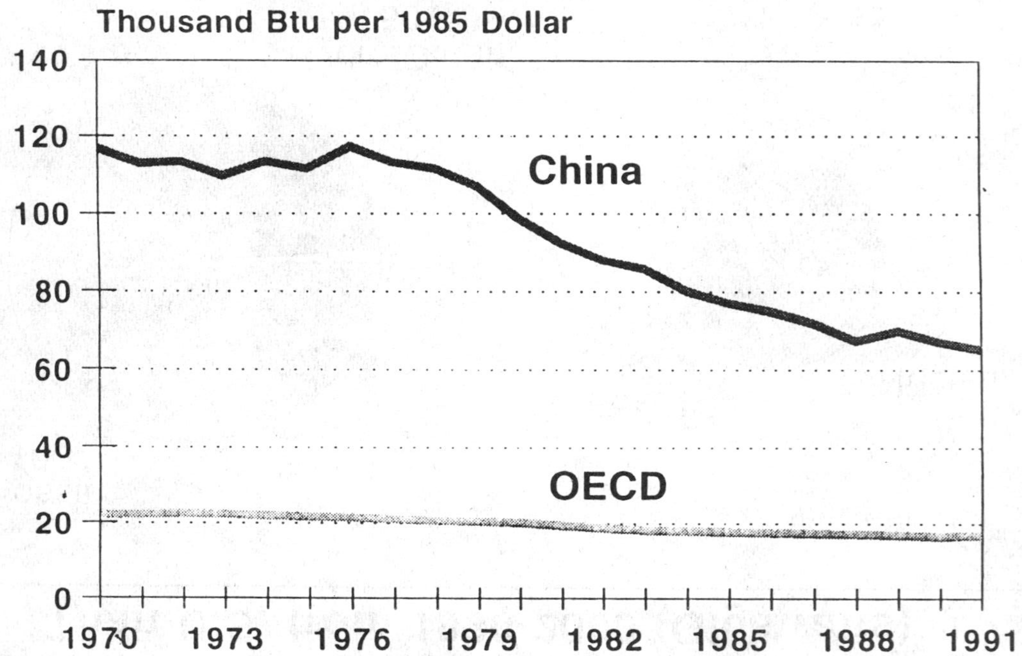

China Adds Electric Capacity Faster Than U.S. from 1990-2000

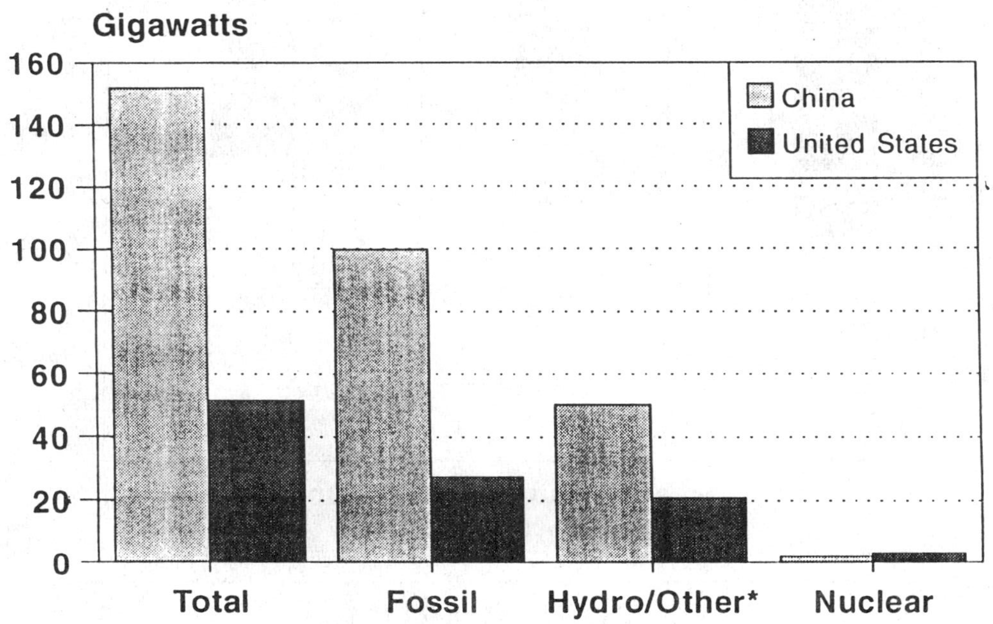

*Includes cogeneration

China's Carbon Emissions Grow Rapidly

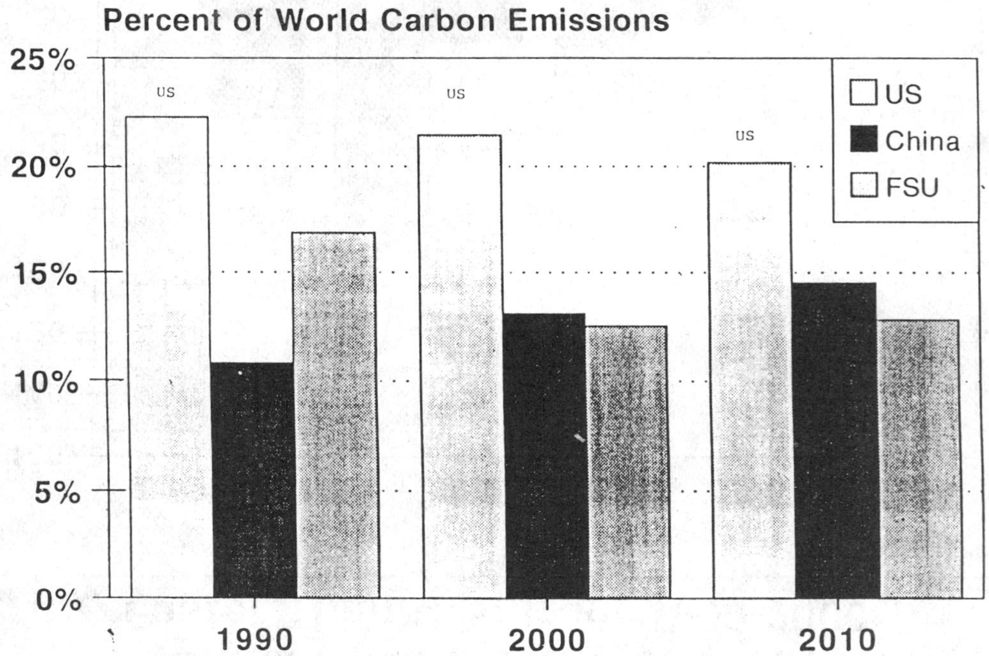

The CHAIRMAN. Senator Bennett, why do we not hear from Mary Nichols, and then we will come back for questioning, in spite of the fact that I violated that rule.

Ms. Nichols is Assistant Administrator for Air and Radiation for EPA, and we are glad to have you here, Ms. Nichols.

STATEMENT OF MARY NICHOLS, ASSISTANT ADMINISTRATOR FOR AIR AND RADIATION, ENVIRONMENTAL PROTECTION AGENCY

Ms. NICHOLS. Thank you, Mr. Chairman, and thank you for the opportunity to address this committee regarding the domestic and international implications of energy demand growth in China. I recently had the opportunity to also lead a delegation to meet with China's National Environmental Protection Agency, and as a result of that I have gained a personal appreciation for the immense scope of China's economic development and the related increases in energy demand, as well as the impacts of this growth on the environment.

Increased energy demand in the Asia-Pacific region will have tremendous local and global environmental impacts. I will not try to relay all the information that is on the charts that you already have, but as you know China's primary domestic energy source is coal. Mining as well as power generation causes severe local air pollution, water pollution, and mine waste disposal problems at mining and power generation locations within China.

Additional air pollution is created through the construction and use of coal transportation networks, primarily rail and truck. Not only does the burning of coal play a major role in China's contribution to global warming, but the release of methane during coal mining also represents a tremendous loss of a clean energy source.

Energy use in China is growing rapidly. In the future, China is projected to become the world's single largest emitter of greenhouse gasses, and as has been recognized previously, if all the developed nations are successful in stabilizing emissions of greenhouse gases and China continues to develop on its current path, the U.S. projects that by the year 2020 China will account for more than one-third of the emissions from all developing countries, and nearly one-fifth of the entire world's emissions.

The environmental impact of this projected growth in energy demand and related greenhouse gas emissions is enormous. Global stabilization of greenhouse gas concentrations is not possible without major reductions in China's emissions growth, even if the rest of the world completely stabilizes their emissions.

EPA's work regarding China is based on our activities to implement the Clean Air Act, as well as to support U.S. participation in the Montreal Protocol, and to support U.S. participation in the Framework Convention on Climate Change. In these areas and in our cooperative research program with the Chinese environmental protection agency, we work closely with other Federal agencies. Section 603 of the Clean Air Act amendments of 1990 requires EPA to report on methane emissions from countries other than the United States, and to analyze the potential for preventing an increase in the atmospheric concentrations of methane from activities and sources in other countries. Reports to Congress to fulfill this requirement were published in 1993 and 1994. The reports, entitled "Options for Reducing Methane Emissions Internationally" and "International Anthropogenic Methane Emissions", identify and evaluate the technical options and costs for reducing methane emissions associated with natural gas extraction, storage, transportation, distribution, and use, as well as coal extraction, transportation, distribution, storage, and use, and include coal bed methane recovery, management of solid waste, agriculture including rice and livestock production, and other emissions associated with human activities. These reports also identify emissions reductions that would be necessary to prevent increasing atmospheric concentrations of methane and identify technology transfer programs that could promote methane emissions reductions in less developed countries.

We recognize that rice production, coal production, and animal husbandry are the largest sources of methane emissions in China today, but of these three coal production has the greatest potential for profitable near-term reductions. In fact, China has the greatest potential of any country in the world to profitably capture methane currently wasted during coal mining. We estimate that coalbed methane could reduce China's need for coal burning in powerplants by about 25 percent by the year 2020. Furthermore, the development of a coalbed methane industry in China is a tremendous opportunity for U.S. business, as the United States is the world leader in this field.

EPA is currently working on a final report to Congress which will present "options outlining measures that could be implemented to stop or reduce the growth in atmospheric concentrations of methane from sources within the United States . . . and to identify programs of the United States and international lending agencies that could be used to induce lesser developed countries to undertake measures that will reduce methane emissions and the resource needs of such programs." This report will identify potential programs to accelerate the transfer of U.S. technology related to reducing methane emissions, and is expected to be submitted by late 1995.

As a result of our investigations in support of these reports to Congress, EPA was able to leverage $10 million from the World Bank, the U.N. Development Program, and the U.N. Environmental Program Global Environmental Facility to support demonstration of coalbed methane as an important source of clean, economically preferable energy for China.

The United States and other developed country parties also have a commitment under the Framework Convention on Climate Change to provide financial and technical assistance to developing countries to enable them to comply with some of the obligations under the convention. In response to this mandate, EPA participates in the interagency country study management team composed of representatives from EPA, Department of Energy, USAID, NOAA, the Departments of Commerce, State, and the USDA. The team provides technical assistance to developing countries on climate change. The country study program helps developing countries prepare emissions inventories, assess vulnerability to climate change impacts, and develop mitigation strategies. A team with representatives from EPA, DOE, and NOAA, as well as the Department of Commerce is visiting China this week in fact to develop details of China's country study. EPA has also worked on developing emissions inventory methods and developed guidelines for reporting procedures and for the review of national plans to prevent climate change.

EPA has worked with several Chinese institutions to assess the impact of climate change and management practices reducing emissions. Typical examples include efforts to understand how climate change will affect Chinese agriculture, the coastal zone areas, and human mortality rates.

And we were informed in China that lung cancer is now the leading cause of death in China, and that is related directly to breathing coal particulate and other constituents.

We have assessed changes in water management practices which could reduce methane emissions in rice paddies while maintaining productivity. Our goal in all these activities is to help the Chinese government understand how it will be affected by climate change, and how it can contribute to possible solutions.

As part of the U.S. contribution to the implementation of the Montreal Protocol, the United States has been assisting China in its efforts to phase out CFC's and halons. These ozone protection activities have also served to reduce emission of greenhouse gases, and provide an opportunity to consider other potential climate change activities. As a result of this work we have identified a number of cost-effective opportunities to work with China to greatly reduce greenhouse gas emissions which arise due to its increase in energy demand.

China and the nations of the Pacific Rim, as this committee has already recognized, offer many opportunities for the U.S. export of energy and environmental technology. As noted in the administration's environmental technologies exports report, the Pacific Rim and China in particular will be one of our focal areas for environmental technology exports. The Chinese government plans to spend over $35 billion by the year 2000 on environmental projects. The United States currently, according to 1991 figures, holds 20 percent of the Chinese environmental technology import market and is working actively to increase that share.

One program that will help disseminate U.S. environmental technology internationally to growing markets such as China is the President's Environmental Technology Initiative, which is led by the EPA. The ETI is predominantly a domestic initiative intended to strengthen the capacity for environmental technology innovation here in the United States, but it recognizes that a vibrant and innovative domestic sector is needed to create technologies that will also win a leading share of markets abroad. One component of the ETI program is called U.S. Technology for International Environmental Solutions, or U.S. TIES. U.S. TIES encompasses a range of activities including international technical assistance and training, information dissemination, financial assistance, technology demonstrations, and efforts to promote the international adoption of regulations and standards. This program will help create additional markets for U.S. exports, and will help protect the global environment. This year the ETI will form an international technology diffusion subcommittee that will include interagency representation. EPA will be working cooperatively with the Departments of Commerce, Energy, Defense, and others to fund those projects with the greatest potential to spread the use of innovative environmental technology at home and abroad.

There are a number of impediments to the U.S. efforts to work with China to reduce its emissions of greenhouse gases. One impediment to U.S. efforts to reduce China's greenhouse gas emissions enough to stabilize atmospheric concentrations is the limited foreign assistance dollars from all countries. U.S. assistance funds should therefore used in a strategic manner to catalyze transformation of key sectors to more energy efficient technologies, to leverage maximum adoption of energy efficiency, to promote clean energy sources such as coalbed methane, conventional gas, and renewables, and to promote high efficiency technologies that greatly reduce the environmental impact of coal combustion. In particular, work should continue in the promising areas of domestic refrigeration, coal bed methane recovery and use, and important opportunities in the areas of lighting and industrial energy use should be pursued. Coalbed methane and the expansion of renewable and alternative energy are two central energy supply areas for concentrated effort. Strategic U.S. assistance to China is a critical step in fulfilling our shared objectives for greenhouse gas reduction under the Framework Convention on Climate Change.

This concludes my prepared remarks, and I am happy to answer any questions you may have.

The CHAIRMAN. Thank you very much. Under the Rio Treaty, if we achieve every goal that we set out to achieve, which I think is quite doubtful, but if we achieve every one of those goals the amount of greenhouse gases to be put out by China would more than overwhelm that. Is that not correct?

Ms. NICHOLS. That is correct.

The CHAIRMAN. And we also understand, as listed in our report, that the potential market for environmental technologies by the year 2000 should be $590 billion. Is that about right?

Ms. NICHOLS. I cannot verify the figure, but it sounds plausible.

The CHAIRMAN. Well, that is from the Department of Commerce in a report entitled, "Toward a National Exports Strategy."

So, here we have this tremendous need, this tremendous market, and yet the United States does not have OPIC available, not even for environmental technologies. We do not have TDA, Trade and Development Agency. We have restrictions on Exim Bank. We are not doing joint implementation under the Rio Treaty.

What is wrong? Where is the failure here, Mr. White? I mean, is this not a clear failure, and who ought to be looking at that?

Mr. WHITE. What is wrong is that the massacre that occurred in China shocked the conscience of a number of people in this country, and the fact of the matter is people historically try to find whatever points of leverage that they may have.

I will not get into the debate that you are very familiar with concerning whether sanctions hurt U.S. citizens more than others upon whom they are imposed. Certainly it is in the interest of American businesses to remind the Chinese of the importance of these concerns to the U.S. Government.

The CHAIRMAN. I mean, it is not just trade. Here the Chinese are belching out greenhouse gases at a record rate and at a record rate of acceleration, and we profess to be very concerned about the world's environment. I hope we are. And yet we say because of Tiananmen Square we are not going to let you clean up the world's greenhouse gases. Is there not a disconnect there somewhere?

Mr. WHITE. You are obviously talking about, whether it is done explicitly or not, a balancing of competing interests. And in all likelihood at the time those restrictions were enacted the magnitude of the competing interests were not appreciated, and I will leave it at that.

I do believe that Assistant Secretary Lord has made a recommendation, and I think that action on that recommendation may be—and I cannot tell you for sure—may be contingent upon the overall White House review that it is undergoing right now.

The CHAIRMAN. What I am saying is that action on the world's environment ought to be separate from actions on America's trade interest. Now, I happen to believe that America's trade interest is connected to America's economy, to America's jobs, to the health and welfare of our own citizens in a very direct and important way, and for that reason I think you ought to delink MFN.

But if you disagree with that, surely it seems to me that people should not disagree with the fact that we should not restrict the export of environmental technologies to China.

Our next witnesses will be talking about some case studies about how bad it is over there. If you have been there lately you will know. I mean, nobody would dare to go jogging out at Beijing. I mean, the air is just terrible.

And is it not in the world's interest to clean that up? And if it is, should we not set aside the environment and should we not have somebody in charge to say, well, let us work on the world's environment by giving them clean coal technologies, because they are going to be burning coal?

And maybe somebody ought to even look at nuclear technologies. After all, they are going to go nuclear anyway. We have the world's safest. How about that, Ms. Nichols?

Ms. NICHOLS. Well, Mr. Chairman, I was just going to comment that one thing that is somewhat encouraging, in light of the bleak situation that you have just described is that environmental cooperation has continued between the United States and China over the past 14 years, regardless of the ups and downs of our political relationship. And certainly the interest in environmental protection is growing in China.

I think there is a great commitment on the part of that nation's leadership and great desire for U.S. technologies. And to the extent that we can either provide them with information or with access to those technologies, there is already a growing demand over there that the measures that you are taking would obviously make a difference.

The CHAIRMAN. Well, that is right. I mean, you talk about cooperation and demand, but under the Clean Air Act EPA cannot give them any technologies. I mean, you can study their problem. OPIC cannot give them insurance. If you have an American company that wants to give them scrubbers or clean coal technology or what have you, the mood is good, the demand is great, cooperation is okay, but where is the trade? Look at those figures over there.

Ms. NICHOLS. The funds are extremely limited, I certainly would agree. But even with those very limited funds, for example, we are allowed under the Clean Air Act to do some small grants work. And, for example, in the next fiscal year we are proposing a flue gas desulfurization project with China's EPA, which will allow them to demonstrate the use of technology that we use here in the United States for desulfurization of coal in powerplants.

The CHAIRMAN. How big is that project? Is that not a $10 million project?

Ms. NICHOLS. Yes, it is small.

The CHAIRMAN. That is minuscule and, I mean, if there is an environmental program anywhere in the world that affects the world's greenhouse gases it is here, and so we have a $10 million program. We put many, many billions of dollars wringing out that last little increment of pollution from American industries, and billions are spent on that, and yet here is our target of opportunity and we let it fly away.

Mr. White, how can we get a national policy that sort of brings this issue up to this environmental administration? I mean, the world environment is a big goal of this administration. We talk about it all the time.

Mr. WHITE. I think this is a very important year for China policy, and I can address, for example, changes that I heard about in the attitudes of some on Capitol Hill when they are confronting the reality of just this issue as well as the trade deficit that we have.

So, I think hearings such as this are important, and as we go through—you know, for better or worse this MFN issue is going to focus commentators, focus the President, and I think how the balance is tipped there and the whole related set of China issues that will ultimately be focused on by the White House will be considered—this is a very important year for China, and exactly the considerations you are identifying right here need to be a big part of that.

The CHAIRMAN. Well, what I am saying is—I mean, MFN has a life of its own. It has its proponents, its opponents, and that issue will be argued and I hope successfully so. But these issues ought to be separate from that.

I mean, if MFN is renewed, then you ought to have an automatic OPIC, an automatic TDA, and automatic Exim Bank, because all of the reason not to do those things would then be gone.

Mr. WHITE. Right.

The CHAIRMAN. And I fear that the administration does not view it that way, or I fear that it will not happen that way, should I say. And even beyond that, if you deny China MFN and if you say that they are not making progress on human rights, therefore we ought to punish them through denying trade, then you ought to make an exception for the environment.

And somebody ought to have an environmental package for China, and indeed for the whole Pacific Rim but particularly for China, that says this is a great target of opportunity and we need to work on this.

There is a press report today that says that $11.5 billion is lost annually in China due to the cost of energy inefficiency and environmental damage, 18 percent of the Chinese Federal budget. I mean, here it is to be done. It is in China's interest, of course. It is in our interest.

[Recess.]

The CHAIRMAN. The hearing will come to order. It is clear that most of the focus of both DOE and EPA has been on China. However, I recently received a letter from the U.S. Ambassador to Indonesia, Bob Barry, urging greater commitment from DOE to work with Indonesia as it develops plans to expand in the nuclear renewables and clean coal technologies.

Has there been any focus on Indonesia?

Mr. WHITE. Yes. As a result of a trade mission that occurred last year, there are two projects there that are just coming to fruition which are examples of what we ought to do more of. One is a gas-fired independent powerplant that will be undertaken by Mobil and Enron, and another is a coal- fired plant being done by Mission Energy and General Electric.

The latter contract has now been signed. The former is close to being signed. We are going to have two trade missions with Indonesia this year. The TPCC has targeted Indonesia as a test case of whether this Trade Policy Coordinating Committee is going to work, because it is such a major market.

And finally, we are very active there on renewables. As an example of that, we have provided the technical assistance in the CORECT program in order to get a photovoltaic project of some $20 million, and now that project has all been spec'd out and is being sponsored by the Indonesian Government for international financing. And that would be using U.S. photovoltaic firms.

The CHAIRMAN. By the way, I understand that in 1993 the following four categories accounted for 76 percent of China's imports. Machinery, mechanical and electrical was 34 percent. Base metals and tools were 17 percent. Minerals, chemicals and plastics were 17 percent. And transportation, including airplanes, was 8 percent.

All of these categories are categories in which the United States is preeminent, or at least competitive, and it just seems to me that those are things that ought to be pursued.

Senator WALLOP. Mr. Chairman, let me make two observations. One, all these nice things that we have in this country that can make other people energy efficient may fall on the alter of the Most Favored Nation trading status. It seems ultimately that that may not be in the interest of the people that we are trying to help.

That is a topic for maybe another committee. I just wanted to express my anxiety about what appears to be a casual, even flippant way of going about making trade policy. I hope somehow or another that your Agency and good offices can kind of keep it in perspective.

I have a friend who happens to be the chairman of British G.E., who told me that he fervently and passionately hoped that we would deny them Most Favored Nation trading status because British G.E. has a very nice set of turbines that they wish to—they cannot compete with us otherwise, but they certainly could if we are not a competitor at all.

Mr. WHITE. We hear that message from you and the chairman.

The CHAIRMAN. Well, thank you very much, and I hope, Mr. White, that Energy will take to heart some of the things that we have said here. Somebody needs to be pushing these technologies within the administration, in particular the environmental technologies, and separate those. If you cannot separate anything else, those at least ought to be separated.

Thank you very much.

Mr. WHITE. Thank you.

The CHAIRMAN. Next we will hear from a panel with Mr. John B. Wing, who is chairman of the Wing Group, headquartered in Aspen, Colorado. Mr. Mark Brody, who is president of the China Environmental Fund in Mount Horeb, Wisconsin. He is accompanied by Mr. William Eldred and by Mr. William U. Chandler, who is senior scientist at Battelle Labs in Washington, D.C.

Senator WALLOP. Mr. Chairman, if one of the witnesses could perhaps explain to us where the mountain was in Wisconsin?

The CHAIRMAN. You mean Mt. Horeb.

Mr. BRODY. That is west of Madison by a half-hour.

Senator WALLOP. Does it look down on Madison?

Mr. BRODY. It is at higher altitude, so I suppose it does.

The CHAIRMAN. Let us begin with Mr. Wing.

STATEMENT OF JOHN B. WING, CHAIRMAN, THE WING GROUP, ASPEN, CO

Mr. WING. Thank you, Mr. Chairman. I do have some written comments that I would like placed in the record.

The CHAIRMAN. Without objection your written statement will be placed in the record.

Mr. WING. I would like to start by saying that we thank you, Mr. Chairman, for being what we think is the most interested and informed member of our Government on what is going on in China and other parts of Asia, and we hope that we can continue to make progress in getting others here on the Hill and in the administration up to speed.

One other thanks. We have all been at this—probably the two of you there and I may be the last remaining standing folks that were involved in the early PURPA discussions, and what we thought was going to be a cottage industry confined to the United States turned out to change how we make electricity and move energy around the United States.

I am not prepared to take credit for the next part, what happened internationally when 60 percent of the world decided to privatize based on what happened in the eighties. And as a result of that we, as Americans, because of what we did with PURPA and everything following that, as a country are more prepared to create this independent power business than any other country in the world.

And that is not just a technological game or a financial game. It is that we have learned how to create these businesses, and many of us owe you a debt of gratitude on that. So, thank you very much.

On a personal note, I have not seen Senator Wallop for awhile. The last time we were together I believe was late at night in Northern England, and at that point he said to me that we should get out of this racket before we are too old to see well enough to tie a fly or too feeble to throw it in the center of a stream. And I understand that at least one of us is taking that advice, and I respect you for it and envy you. And if the invitation is still there to come up I want to do that.

Senator WALLOP. It is by all means, and you would enjoy it.

Mr. WING. I will talk briefly about Vietnam, and then a little more detail regarding China.

My comments on Vietnam are very simple. I have a little bit of a personal stake there. I was a pilot there. When I go out for my morning trots here in Washington, I go by the monument where the names of three of my roommates and best friends are placed.

And I want to say that I believe that what the President is doing in lifting the sanctions is absolutely 100 percent the correct thing to do. And I also believe that that shows a significant amount of courage on his part to get that behind us. I think the trip that you all made there not only helped in that process but probably was an eye-opening experience for you.

There is not a better way, in my estimation, to get 30 years of awkwardness and pain behind us or a better tribute to those involved than to get on with the business of what we were trying to achieve when we first went there, which was to make things better for that country. So, I think he is on the right track. I appreciate your help in moving that forward.

I do think the next step—I tend to come at the world not from how things look up on Mt. Olympus, but kind of the dirty fingernails end of it. And I do think that the next step is a Government-business taskforce to look specifically at what we can do to rebuild infrastructure in Vietnam, I urge also Cambodia at this point, to try to get that chapter closed.

I think that I am very conscious of the issues veterans and other groups have. Yes, we did lose more than 50,000 people. They lost more than 500,000. And the time is now to get on with it and I think it is a show of statesmanship to do that.

If you look at that 30-year followup and you kind of look at where we are in China, one might easily conclude that the problem that we got ourselves into was a problem of not defining what we wanted as an outcome in Vietnam. I am not sure that we ever did that in the 10 years that we were actively involved. You can look at that and compare it with what went on in Desert Storm, where we clearly articulated what we wanted to achieve.

And I would urge the Government, when it is looking at China, to make very sure that we do not go ready, fire, aim in terms of our policy with them. Foreign policy experimentation, when you are dealing with the size country that China is, the size of its problems, the size of its economy, is not something that we should take on.

The dilemma with China from the businessman's perspective cannot be isolated totally to what goes on on the four corners of a page of a contract with the Chinese. I believe you have to take the economic issues, the strategic issues, the environmental and the human rights issues all together. And I would like to just do that quickly, if I might, and then come out with some recommendations.

The economic issues should be pretty clear. There has been a lot of talk by some pretty senior people around this town regarding jobs. This economy will be the biggest in the 21st century in our specific industry. It will be the biggest opportunity of our lifetimes.

We at the Wing Group have somewhere between $14 and $15 billion of letters of intent in China. Our Government says each billion is about 20,000 jobs. Roughly 70 percent of those would be in the United States if all goes well.

But let me be very clear with the committee that this is, in fact, a live fire exercise. We cannot sit and debate Most Favored Nations forever. Currently, my principal suppliers are Westinghouse, and G.E., and Bechtel, and Fluor, and Babcock-Wilcox, and Riley, and they are in my estimation the most highly qualified to provide the goods and services. My financial advisors are for the most those found a couple hundred miles to the north of here.

If we do not grant Most Favored Nation, so that we do not have a dispute about who said what, we will change those. Instead of Fluor and Bechtel it will be Hultsman and Marabini. Instead of Westinghouse and General Electric it will be Mitsubishi and ASEA-Brown-Boveri and Siemens.

And instead of those jobs—I am not going to debate with you whether it is 200,000 or 300,000 from our work. But instead of those jobs being American they will be German and French and Swedish and Japanese and Korean.

So, again, back to the starting point. What is the desired outcome?

Let us focus on what the mission is and what the objective is, and not get ourselves tangled in our underwear worrying about the bumps in the road along the way. The strategic issues also I believe are a function of the economic. We are not going to be able to work effectively in their economy if we are absolutely at each other throats on strategic issues.

Frankly, we cannot control a lot of that, but my advice to the Government is the following. Put yourself 25 or 30 years out. Now, we all are doing what Malcolm is contemplating, and we are looking back from there at today, and what should we have done?

Option one is a strategic situation where we are totally separate from China, which then I believe would align with India, Southeast Asia, and Japan.

Option two is that there is an interdependence between the United States and China. In the short term, if we think that nuclear issues in North Korea are important, then clearly being friends with China is important.

In the longer term, which of those two options would you prefer? Again, the outcome that I prefer is that we be interdependent.

The environmental issue we have talked a lot about this morning. Again, I will not rattle through a bunch of numbers because I do not think that this is a numbers issue, but I believe that the greatest human right that we can give the Chinese people, if we can do anything at all right now, is the right to cleaner air and clearer water. And when we take restrictive positions—let us be really clear with one another. The bottled water that the leaders drink will not change. That will be fine.

Restrictive positions are not going to hurt them. I have not seen the leaders of the country on bicycles with masks because the air is so dirty going through Beijing. They are not the ones who get hurt.

So, again, what is the outcome? My sense is these restrictive positions isolate us and isolate the people that I would think we are trying to help. And I think indeed you would get some interesting perspectives.

Even if you were to ask the dissident groups in China whether or not they believe the United States being involved in the economy and not using Most Favored Nations as a political stick is a good or bad idea, I think people would be surprised.

On the human rights issue, just to summarize quickly, I believe that we need to find a way to get off our high horse and to accept the fact that the rest of the world does not totally need or want an Anglo-Saxon capitalistic structure, and no matter how long we hold our breath that is not going to happen.

I believe that there are things that need to be improved, and they come in all directions. They do not put absolute, total, individual rights at the base of their human rights pyramid. In fact they refer to them as corrosive individualism. They put food, clothing, and shelter as the basis what they need.

I am not sure that we have the moral high ground to say that that is wrong. They comment on our 25,000 murders a year and they comment on our support of other countries. It is a little unclear to them how someone blowing a hole through the Russian White House is a step for democracy and what they are doing is so terrible.

So, I think a little bit of tolerance, a little less hot rhetoric starting from the position that we do not always have the total right answer. Listening would not hurt. If we want to get at the human rights issues, I believe that the right way to do it is through the economic and environmental issues.

They are doing a lot. You know, they are feeding 1,200,000,000 people. All is not perfect, I will admit that, but it was only in 1959 and 1960 when Chairman Mao turned left instead of right on his agrarian policy that they starved 30 million people. There has been quite a bit of progress since then. They have a different structure. They will have, no matter what we do.

We do not need to compromise our values, what we want to get out of it. We need to compromise in how we deal with them in order to achieve our values.

That is really the end of my comments, Senator. I will be happy to answer any questions.

[The prepared statement of Mr. Wing follows:]

PREPARED STATEMENT OF JOHN B. WING, CHAIRMAN, THE WING GROUP

Chairman Johnston and members of the Committee on Energy and Natural Resources, thank you for the opportunity to testify on the domestic and international implications of energy demand growth in China and the developing countries of the Pacific Rim. Before I begin my testimony, I would like to express my appreciation to the Chairman for the significant leadership role he has taken on the critical issue of U.S.-China relations and for the time and effort he has committed in particular to advancing U.S. interests as China develops its energy infrastructure. I also would like to commend Secretary of Energy O'Leary, Deputy Secretary

White and the Department of Energy for their constructive efforts to promote the use of U.S. electric power technology and services in China and around the world.

The Wing Group is an internationally recognized leader in the field of independent power development with projects under development in China, Turkey and Kuwait. We are currently involved in power projects in China totalling 14,600 megawatts (MW), which including liquefied natural gas (LNG) facilities represent a total possible investment of $19.6 billion. These projects in China include natural gas, coal and wind technologies.

Because of our experience in that country, my testimony today will focus primarily on China. Nevertheless, much of this testimony applies equally to the other developing countries of the Pacific Rim. I also will address recent developments in Vietnam. Finally, I will conclude with specific recommendations for needed changes in U.S. policies affecting our commercial relationships with China.

CHINA

To look at China today, one must integrate the geopolitical, the economic, and the environmental. Economically, the questions are very simple. Is economic interdependence with China desirable in the twenty-first century? Is it something that we should work hard to achieve? Should we be neutral or negative? Our views are that economic interdependence with China is a must if we are to remain a world power economically. Such interdependence is critical not only to our own economic interests, but also to our strategic and environmental interests.

In the energy sector, the opportunities in the China power market are enormous—perhaps greater than anything we have or will see in our lifetimes. While China already ranks fourth in the world in terms of total generation capacity and has the largest generating system of any developing country, power shortages and rationing are commonplace at present. In fact, shortages of electricity force many ventures in some parts of China to shut down during certain hours, or even for days at a time. In addition, about 120 million people in rural areas—equivalent to about half the U.S. population—do not have access to electricity. In order to address these power shortages and to meet electricity demand growth projected to be at least 10% annually, China plans to almost double its electric generating capacity over the next decade, adding over 150,000 MW. In a major departure from decades of central planning, China is making a significant part of this power development available to the private sector and is welcoming foreign private sector participation.

This represents a tremendous market for U.S. manufacturers of generating and environmental control equipment, as well as U.S. construction contractors, developers and plant operators. Even assuming that China is able to produce domestically half of the equipment and services needed to meet these capacity additions, there is a $75 billion export market at stake here over the next decade. The Department of Commerce estimates that each $1 billion in exports creates 20,000 jobs and increases real Gross Domestic Product (GDP) by twice the amount of export value. Industrialized countries thus will be competing for about 1.5 million new jobs and a $150 billion increase in GDP. It should be our mutual goal to make those U.S. jobs and U.S. economic growth.

While the exact size of the U.S. share of the future China power market will depend on the actions of both the U.S. government and the private sector and is thus uncertain, two things are certain. First, the potential domestic economic impact of our relations with China are significant. These are real, high-paying, high-technology jobs, many of which will be in the manufacturing sector. Moreover, these jobs and exports translate into real economic growth and also directly impact the U.S. trade balance and the Federal budget deficit through increases in tax revenues and reductions of government spending for unemployment and welfare benefits. Second, China needs the power and if U.S. policymakers impede our ability to compete for these projects, China will still build these plants but they will do so either alone or with our international competitors. In other words, we will deny a U.S. worker a job and give it to a worker in Japan or Germany or Sweden or Korea. We will replace the Bechtels and Fluors with Holtzmans and Morebenis. We will replace the Westinghouses and General Electrics with Seimens and ABBs and, yes, we will replace Wall Street with the City of London or with Hong Kong.

It is important to remember that our actions and decisions today will have an impact not just on the projects currently under development, but also over the long term. If American firms dominate the first round, our win will carry over the next three decades. But if we are shut out now, we will be at a severe disadvantage in future rounds. These plants have thirty to fifty year lives; reputation and performance count heavily. We must be reliable partners, offering the best, most competitive and most environmentally benign product. If given the opportunity, I am confident that U.S. companies will do just that—bringing economic benefits to China and the U.S. and environmental benefits to the region and the globe.

Unfortunately, the environmental impacts of China's growth and the potential U.S. role in mitigating such impacts are often overlooked in the rhetoric of the China debate. The rapid growth of energy demand in China has significant implications for local air pollution, regional acid rain problems, and global warming. China is the world's largest producer and consumer of coal. Coal will continue to be China's main source of energy for the foreseeable future and is projected to maintain approximately 70 percent of the electricity market to 2010. As the Committee is aware and as others will testify today, China's expanding coal use creates serious regional and local environmental problems and has important ramifications for human health and for global efforts to address climate change. Indeed, projected increases in China's carbon dioxide emissions could more than offset efforts of the United States and other countries to stabilize their emissions at 1990 levels.

The Chinese are making progress in addressing environmental issues, and we need to work with them as partners to achieve global environmental goals. Otherwise, the resources and costs we expend domestically to meet greenhouse gas emissions levels could be severely undercut. U.S. technology and services can play a critical role in reducing the global environmental impacts of China's power development, while simultaneously improving the local and regional environment and the health of the Chinese people.

The U.S. is the world leader in environmentally sound energy technologies. Our technology can help China increase its energy efficiency, utilize its vast coal resources more cleanly and efficiently, and look to other sources such as natural gas and wind power. As noted earlier, our current efforts in China involve coal, natural gas and wind technologies. There is clearly a great need for high efficiency, environmentally preferred U.S. clean coal technologies, and it is projected that China will become the largest user of such technologies in Asia. China's needs are particularly suited to the large scale clean coal technologies developed in the U.S.

In addition, we should lead efforts to develop natural gas and renewable energy electricity generation. Let me give you an example on how we in our group develop environmentally sound power projects. A consortium of companies we organized recently completed a large power plant in the United Kingdom called Teesside. Teesside is the largest independent power plant in the world. It utilizes natural gas, it is very efficient and it is very environmentally friendly. In fact, the construction of this plant allowed the UK to comply with European Community air quality standards; had it not been built they had no way, even looking at all sectors combined, of complying with those standards. When we went there everyone laughed because we wanted to build a gas fired plant. It was not the way they had done things before; coal and nuclear were the answers. But we did it and we achieved this environmental victory. At the same time, the power from this plant is nearly forty percent less costly to the end user than the power produced previously by government owned entities.

We believe in building the least cost, most reliable, most environmentally benign solutions to infrastructure problems around the world. And we are the best in the world at it. Do not let U.S. policy stop us from doing this in the biggest market in the world. I challenge anyone in this room or any environmental organization to duplicate the environmental, much less the economic, outcome that we can achieve by taking what we know how to do so well as Americans and doing it in other parts of the world. Parenthetically, you will be happy to know that all the American companies involved in Teesside made money, paid taxes and significantly helped the balance of trade. And again, let me say, that the greatest opportunity we have in this area, in the world, and will have in our business lifetimes, is in China. Good economic solutions and good environmental solutions need not be mutually exclusive. They won't be in China if our own government does not stand in the way.

VIETNAM

Next, I would like to quickly address recent developments in Vietnam. While there is a tremendous need to improve the inadequate Vietnamese infrastructure, including energy supply, I would like to comment on the much broader issues of U.S.-Vietnam relations. Several of you have asked my views on the President's lifting of the sanctions on Vietnam. As you may know, I am a Vietnam veteran, was a pilot there, and had many friends who were killed, missing or POWs. So, I have some knowledge as well as a personal stake in the situation. I believe it is a clear act of leadership by President Clinton to lift the sanctions and get that behind us. He is doing the right thing. He is taking the first major step in fixing a mess none of us can be proud of. We do not need further debate about his past views or actions.

82-926 0 - 94 - 2

His present actions, properly followed up upon, can help right a huge, longstanding wrong.

If President Clinton were to ask me to work on a government/business task force in Vietnam to rebuild infrastructure, to get businesses going by getting Vietnamese and Americans working together to heal past wounds by helping both economies, I would jump to volunteer. I hope the groups opposing the President's recent steps think about this a little bit harder and help President Clinton become the President who turns a three decades old failure into something good for both countries and their people. They need us; this time let us do it right. Helping the Vietnamese people and making both countries stronger and better is what we wanted to do in the first place. What better tribute to those who gave the ultimate sacrifice than to complete what we set out to do. Not only veterans, but all Americans should support the President in trying to finally make something good come out of our role in Vietnam. I support him wholeheartedly on this effort.

RECOMMENDATIONS

Before ending with specific recommendations for needed changes in U.S. policies affecting our commercial relationships with China, I want to return to the first point I made which is that our strategic, economic and environmental goals must be integrated. While we have political differences on issues such as human rights and weapons proliferation, economic and environmental cooperation in electric power makes sense for both China and the U.S. and can help further not only our economic and environmental interests but also our strategic interests. Conversely, if we allow our political differences to impede or even destroy these opportunities for cooperation in electric power, then we will harm our strategic, economic and environmental interests and the ultimate losers will be both the Chinese and the American people.

I firmly believe that economic interdependence and constructive engagement are the best levers for promoting our interest in China. Environmentally sound U.S. power technologies can directly improve the quality of life of the Chinese people by providing electricity to those that currently have little or no electricity and by reducing the environmental and health effects of China's current energy consumption patterns. As a key element in China's ability to achieve sustainable economic development, such projects indirectly improve the quality of life and choices available to individual Chinese. The benefits for the U.S. economy and the global environment are clear. Strategically by increasing cooperation in this key sector, we improve U.S.-Chinese relations and increase opportunities to work cooperatively to address not only our differences, but also issues where our interests may converge such as the environment and North Korean nuclear proliferation. Moreover, economic growth and interdependence can serve as powerful mechanisms for democratic reform. The alternative path of confrontation, trade sanctions and unilateral disengagement would destroy the progress made to date on all fronts and unnecessarily harm U.S. interests.

U.S. policies should support commercial relationships with China. Instead, we face some serious Washington impediments to U.S. development and financing of power projects in China. Some specific recommendations follow:

We should extend Most Favored Nation (MFN) trade status without conditions and should enact legislation to make unconditional MFN status for China permanent. MFN status is an overly broad and unwieldy tool for promoting our interests in China. The annual debate over extension is counter productive and diverts resources from more constructive, flexible and targeted foreign policy tools. From the perspective of power development, our ability to develop and finance projects is severely hampered. Project development and financing is a multi-year process and the uncertainty engendered by the annual MFN debate raises questions as to our reliability as partners and puts U.S. companies at a competitive disadvantage.

We should lift the constraints on the use of Overseas Private Investment Corporation (OPIC). Trade and Development Agency (TDA). and Agency for International Development (AID) programs in China. Our international competitors have the advantage of significant government export promotion support without such limitations. Moreover, OPIC loan insurance or guarantee programs often are key elements for attracting other financing and putting together a funding package for these large projects. AID and TDA programs play an important role in laying the groundwork for energy and environmental infrastructure projects.

Encourage the steps being taken by the Export-Import Bank (Eximbank) to take "project risk" in China, as in other countries. Eximbank has reversed previous restrictions and is now accepting applications for limited recourse project financing in China. This needs to be encouraged.

Thank you for the opportunity to testify today. I know my comments are broader than just energy and business. That is on purpose. One cannot, in today's world, separate the geopolitical, the economic or the environmental one from the other to win on any of the three. We, as businessmen, must understand more than the economic. You, as government leaders, must also find practical, sensible ways to advance the three in a workable strategy for our Nation. I hope these comments are of some help. Thank you.

The CHAIRMAN. Thank you very much, Mr. Wing. Next, we will hear from Marc Brody of the China Environmental Fund.

STATEMENT OF MARC BRODY, PRESIDENT, CHINA ENVIRONMENTAL FUND, MT. HOREB, WI, ACCOMPANIED BY WILLIAM ELDRED, CHIEF POLICY ANALYST

Mr. BRODY. Thank you very much, Senator. In my remarks today I would like to also touch upon some comments I heard from people who provided earlier testimony. Overall I would like to say that I feel very fortunate to have witnessed the transformation of Chinese society over the last 12 years. During my numerous visits to China I have been impressed by the rapid progress the country has made, not just in economic terms but in all types of social improvements for the people at a very basic level, as Mr. Wing has just commented upon.

I think Mr. Wing's comments about listening would not hurt is also a good place to start. I have had the opportunity to listen to Chinese environmental institutions over the past year, and I have heard repeatedly their earnest interest and their commitment to improve environmental conditions in China.

The institutions are still in their infancy, and they are competing against market forces that are perhaps the most powerful and rapid market forces in world history. Never before have so many people industrialized so quickly.

And I think not just from a business point of view but, again, even from a human rights point of view it makes sense for us to become constructively engaged in China, to address both global environmental concerns, concerns for U.S. exports, and concerns for the fundamental rights of Chinese people, as Mr. Wing suggests, to clean air and water.

I would also like to suggest that the United States looks at the total world involvement in China today. Ms. Nichols commented that total foreign assistance financing to China is not doing enough. In testimony we have prepared we cite that all sources of environmental financial assistance going to China now account for approximately $1.4 billion annually. Estimated environmental infrastructure totals a need on an annual basis of $40 billion.

China is falling dramatically short of being able to——

The CHAIRMAN. $40 billion in China?

Mr. BRODY. $40 billion annually in China. Excuse me, it is $40 billion in Asia. Excuse me, excuse me.

But it is $1.4 billion by the World Bank, the Asia Development Bank, OPIC and other types of concerns.

So, the question is, how do we address these challenges? And I would like to thank you, Senator, for your pointed remarks in encouraging this administration to reform its policies and permit organizations such as the US-Asia Environmental Partnership, the Trade Development Agency, the Overseas Private Investment Corporation, to provide that type of jump start or catalyst for U.S. industry to get involved in China.

When I listen to the Chinese people about their environmental concerns, repeatedly I hear of what the Japanese are doing, what the French are doing, what the Koreans are doing, the Germans are doing, and too often they ask where are we? And I think it is time that Americans get involved actively in China, and I very much thank you for your remarks today.

I would like to turn over my comments at this time to William Eldred, who is acting on our behalf, doing a review of both environmental conditions and economic conditions for energy in China and in the Shenyang region. The Shenyang region happens to be the fourth largest city in China, the industrial center of the northeast, a population of 6.6 million, and is in desperate need of environmental assistance.

[The prepared statement of Mr. Brody follows:]

PREPARED STATEMENT OF MARC BRODY, PRESIDENT, CHINA ENVIRONMENTAL FUND

PART I

Mr. Chairman and Members of the Committee, thank you for this opportunity to comment on the explosive economic growth we have observed in Shenyang, China, and its impact on the environment and energy demand.

In Part I of our testimony, we will first describe our work in Shenyang and share with the Committee our key findings. We will also discuss the problems faced by Shenyang in the context of the national economy in China, with a particular emphasis on energy. We will then conclude with observations on a critical "financing gap which we see as an obstacle to environmental problem solving in Shenyang, China, and all of Asia.

For the record, we are submitting the *Executive Summary of Findings and Conclusions* of a Environmental Assessment the China Environmental Fund completed this year for the Municipality of Shenyang, China's forth largest city ad the capital of Liaoning Province. This is labeled as Part 1, Exhibit A.*

(The information now summarized is presented in more detail in the Executive Summary and in the final report entitled *Shenyang Environmental Assessment. Phase I: Summary of Existing Information Recommendations for Action.* The full text of the report will be made available to the Committee upon its request.)

In just 60 years, Shenyang has grown from a small city to a major center for manufacturing and trade with a regional population of 6.6 million. The Municipality covers a area the size of the state of Connecticut, and includes a core city, suburbs and rural hinterland. Shenyang's rapid rise to prominence and the character of its industrial base is reminiscent of the City of Chicago in about 1930.

Mr. Chairman, environmental conditions is Shenyang stagger the imagination. The World Bank has identified Shenyang as the most air-polluted city in the world. During the winter season—when a temperature inversion traps soot and sulfur dioxide emissions close to the ground—mid-day looks like dusk. And, in the categories of water, energy, waste, and land use, the difficulties are no less dramatic. Shenyang health officials report that between 1983 and 1990, the mortality rate associated with cancers increased by 25 percent and associated respiratory diseases grew by 255 percent.

Recognizing the urgency of their crisis, the former Mayor, Wu Disheng, invited a team of U.S. environmental professionals in October 1993 to work with Chinese experts on a rapid initial assessment of the Municipality's challenges; and to make some preliminary recommendations for their resolution. Mr. Chairman, I would like now to highlight the key findings of our report:

Land Use—The core city is home to 1.7 million people who live among the 44 most polluting industries. Crowded together in densities of up to 80,000 persons per kilometer, Shenyang residents have less than half the public green lands recommended by national standards.

---

* Retained in committee files.

Transportation—Today there are more vehicles in Shenyang than there were in all of China in 1979. The 200,000 vehicles now on the road cause massive traffic jams, and few of them are equipped with even the most rudimentary emission control devices. By the year 2000, Shenyang will have as many as a million vehicles.

Drinking Water—Shenyang's drinking water is not safe. Ninety-one percent of wells sampled fail to meet national standards for arsenic, cyanide, ammonia-nitrogen, phenols, and oils. Intensive use of groundwater has caused the underlying water table to drop 33 meters, and this decline is proceeding at an annual rate of 1.5 meters per year.

Many drinking water wells are located in or near industrial areas, or draw on groundwater recharged from the highly polluted Hun River—into which untreated domestic sewage and industrial effluent flows. There is currently no wastewater treatment, although one plant is under construction.

Air Quality—According to the World Bank, Shenyang has the world's highest concentration of soot (suspended particulate) and sulfur dioxide pollution. Though all parts of the City have exceeded standards for Sulfur Dioxide, "industrial districts" have reported concentrations 3 to 6 times higher than the U.S. National Ambient Air Quality Standard.

The concentration of Total Suspended Particulates was 9 to 13 times greater than the U.S. standard. About 70 percent of these particulates are small enough to lodge permanently in the human lung, and are regarded by health officials as a leading cause of respiratory disease.

Energy—The operation of the Shenyang economic engine results in the combustion of 1.4 tons of coal for every man, woman, and child in the city. This compares to a national average of .9 tons per person.

Shenyang used 13 million tons of coal equivalent during 1993 to fuel the industrial, domestic, and transportation sectors of the economy. Of this amount, 9.7 million tons was coal. This will grow to 14.5 million tons by 2000 if energy use continues to grow at the current annual rate of 5 to 6 percent and coal maintains its proportional contribution to the mix of fuel sources now used. Analysis of national energy trend data indicates that continued reliance on coal is likely.

Shenyang's two major electrical power plants burn coal and produce a total of 550 MW. The city has plans to increase the capacity of these plants by up to 425 MW and to develop a 600 MW mine-mouth facility at northern coal fields.

Mr. Chairman, this completes our capsule review of conditions in Shenyang. It has become clear to us that the condition of Shenyang's economy and environment is largely driven by the performance of the national economy as a whole. For this reason, I would like to discuss national issues that will have an impact on the City of Shenyang and all of China's cities.

As the Committee knows, the future of China's environment—air quality in particular—will be determined by its energy system. It is important to point out that industry in China has been working under a chronic electricity shortage that has retarded industrial production by as much as 40 percent. This means that the Chinese economy could grow even faster as plans for additional generating capacity come on line. In this connection, the China Business Review reports that China hopes to add 135 GW of electrical generating capacity by the year 2000, or about 17 GW per year. The report says that 17–20 plants may be built each year, most of them coal fired.

The rapid growth of the power sector will have two obvious impacts. First, there will be the primary impacts from the new plants of higher emissions of carbon dioxide and sulfur dioxide (among others). Second, additional impacts will result from new end-users as the economy responds to a greater abundance of power. In short, more power plants will mean more industry, faster growth, and more atmospheric pollution.

It is important to note that the Chinese economy has made great strides in using energy with greater efficiency. The World Bank reports that energy intensity declined form 2.4 KGOE to 1.73 KGOE. But equally important is the point that—although the energy being used is producing more per unit expended —the total amount of energy expended will continue to rise precipitously no matter how efficient the energy system becomes.

In other words, Mr. Chairman, the challenge of responding to the modernizing Chinese economy and the pressures it will place upon on the global environment—particularly the atmosphere—are unparalleled in human experience. According to one published report citing Japanese sources, China has already become the world's number one emitter of Sulfur Dioxide.

This observation has not been lost on the foreign policy community.

Writing in the December 1993 issue of *Foreign Affairs*, Nicholas D. Kristof, former Beijing Bureau Chief for the *New York Times*, makes salient and prescient observations about the future of China and its impact on the world economy and environment.

He cites World Bank statistics predicting that China's GDP will roughly equal that of the United States by the year 2002 (pg. 61) and warns that this robust expansion will eventually require a "dramatically larger share of world economic resources", observing that already—"the Chinese construction boom has caused a global scramble for certain kinds of steel" (pg. 61). And as modernization raises living standards, more households will acquire televisions, refrigerators, and automobiles. If and when the "each Chinese uses as much energy as every Korean does now", Kristof argues, "then China will use more energy than the United States" (pg. 64).

China's emission of greenhouse gases is now growing faster than any other nation. According to The Stockholm Environment Institute, Kristof says an economic growth rate of 8.5 percent will cause China to "produce three times as much carbon dioxide as the United States" by 2025. But Kristof points out that "this is perfectly fair, since China will have far more than three times America's population.

Thus, Mr. Chairman, it is clear that if and when China and its people enjoy the benefits of the modern age in the same way we do here in the United States—the consequences to the global environment will be devastating—unless the character of development in both countries becomes more sustainable. We are not suggesting that China should be denied the right to modernize or that developed nations be asked to roll back 200 years of material progress. Rather, we wish to call attention to the serious environmental problems that unplanned and unmitigated development will cause for all nations, north and south, east and west.

The root of the problem, we believe, is under investment in sustainable development practices. At no time in history have financial resources been more out of step with human need. According to the most optimistic estimates, the World Bank invests no more than 15 percent of its annual loan portfolio in environmental projects: In fact, total world investment in Asian environmental projects—including the combined resources of the World Bank, the Asian Development Bank, the Japanese Overseas Economic Cooperation Fund, the U.S. Agency for International Development, the United Nations Development Program, and the Asian Development Bank—total only 1.4 billion per year. This compares to an estimated need of $40 billion a year in environmental investment through the year 2000. This means that only about 5 percent of Asian environmental needs are being met with resources now at hand.

Mr. Chairman, the source of the information cited above is the World Bank. We want to make the Committee and the American public aware that the world financial commitment to environmental preservation in Asia is very small. It is clear that sustained and effective actions to prevent environmental degradation in Shenyang and in China will require new and innovative approaches to mobilizing investment and directing it to sustainable practices. We agree with the World Bank and other international organizations that a much higher level of private sector investment will be required to fill-in this huge "funding gap".

We hope that the Committee will develop additional testimony on the "environmental funding gap" that we have described before you today. This is a problem that faces Shenyang, China, Asia, and the world.

PART II

Unfortunately for China, the information presented here paints an ominous future full of environmental crises. Despite the gravity of environmental conditions in China, the China Environmental Fund wants to acknowledge that many Chinese agencies and institutions are making progress, while competing against powerful market forces. We are encouraged by the openness and courage of the Chinese government and Chinese people to speak honestly about their environmental problems, and their earnest attempts to improve environmental conditions in the midst of humanity's greatest historical transformation.

Powerful forces are remaking Chinese society and forever changing the vast Chinese landscape: from the blossoming of individual initiative to the profound and rapid liberalization of market forces, China is 5000 year old civilization facing enormous challenges and opportunities. Above all we hope that many Americans will see for themselves what we have seen: a sense of optimism and enthusiasm amongst the Chinese people that is infectious and inspiring. Now is the time for Americans to become constructively engaged in the world's most dynamic and rapid change: the industrialization and urbanization of China.

This portion of our written testimony has three objectives:

1. To introduce and discuss the work of the China Environmental Fund;

2. To broadly characterize the magnitude of change occurring in China;
3. To explain how this change affects the environment in China.

I. History and Description of China Environmental Fund

Mr. Chairman, as Founder and President of China Environmental Fund, I would like to take this opportunity to share some of my personal experience in China. For the past twelve years I have regularly traveled to and worked in China. My arrival to Beijing next month will begin my sixth trip to China during the last 16 months. In September 1993, I had a unique position as the only American to volunteer to join the Chinese delegation in Monte Carlo to await International Olympic Committee's decision on the Host City for the 2000 Summer Olympics. I consider myself fortunate to witness the transformation of Chinese society, and to build productive working relationships with Chinese businesses, institutions, and citizens.

Building upon past experience, I am committed to expand our organization's activities in China. I understand that many Committee Members joined Sen. Johnston during the recent delegation to China, and to support the work of the Committee, CEF is prepared to keep Members and staff of the Committee informed of its activities, and as appropriate, help coordinate future itineraries in China.

Under a separate cover, we will transmit a document describing the mission and the activities of the China Environmental Fund.

II. Characterizing the Magnitude of Change

Trying to characterize the magnitude of change in contemporary China is difficult; the scale and pace of industrialization and urbanization of China is without global historical precedent. Even the closest historical parallels such as the American Industrial Revolution of the late 19th and early 20th Century or the rebuilding of Europe after WWII are many magnitudes smaller and less environmentally significant than current events in China. A comparison of populations of these critical periods demonstrate the numerical magnitude of China's present industrialization and urbanization.

Year	Country	Population in millions
1890	USA	63
1950	Europe	392
1991	China	1,200

Similar to the energy and drama of the California Gold Rush, China is industrializing with 20 times the population the U.S. had when it began its drive to become the world's leading economy a century ago.

Other statistics on the growth of China's economy speak for themselves:

"Today, China is the world's third largest economy with a projected gross domestic product of $2.8 trillion for 1994, according the World Bank . . . many experts predict that China will become the world's biggest economy within the next ten to fifteen years." (Personal Selling Power, Jan/Feb '94)

China GNP increased three times from the beginning of economic reforms in 1978 to 1993. (China Daily 2/28/94)

". . . in 1992, the Eastern coastline from Canton to Peking enjoyed dramatic increases (in industrial production) ranging from 25 to 45 percent." (Personal Selling Power, Jan/Feb '94)

"The U.S. Department of Commerce reports that in 1992 China imported $80 billion worth of goods and services. That's a healthy 28 percent increase from the previous year (of 1991) . . . In 1992 American companies sold $9 billion worth of products and services to China." (ibid.)

"(For January 1994) Foreign-funded enterprises and private firms reported a 102.2 increase in monthly output over January 1993. The 1993 rise was 64.4 per cent." (China Daily 2/28/94)

In Guangdong Province, arguably the world's fastest growing and most significant future economic zone, the numbers are humbling. (All Guangdong data is provided by the U.S. Consulate Guangzhou, Economic Section, 2/14/94)

In 1993, Guangdong accounted for 28.6% of China's total export (up from 24.3% in 1992) and 40% of China's total foreign investment.

(Following numbers are in Billions of RMB) 1993 Guandong's GDP was 314.7, a 21% increase of 1992, and 1992 had a 19.5% increase over 1991, and the average yearly increase from 1978 to 1993 in real terms was 13.8%. Total trade increased

58.7% in 1993 compared to 1992, and the average yearly increase in trade from 1978 to 1993 in real terms was 23.5%.

Since economic reforms began in Guangdong 15 years ago, the transformation from a central planned economy to a market economy is virtually complete. In 1978, the Central Government funded over 80% of investment in Guangdong. In 1992, only 2% of investment came from the Central budget. The following chart shows how the non-state sector of the economy has rapidly displaced the State sector. Additionally, the number of commercial retail transactions conducted at market determined prices has expanded as follows: from none in 1979 in a centrally planned economy to 98% in 1992.

	State sector	Collective sector	Non-state sector
1979	73%	27%	0%
1992	32%	22%	46%

III. HOW ECONOMIC CHANGE IS AFFECTING THE SOCIAL SYSTEM

Market forces are moving so quickly that China is "Struggling for Control" states the Business Weekly headline in October 25, 1993. The article goes on to say, "China's move toward free enterprise has become a free-for-all." A China Daily headline from February 1994 states 'Planners in plea to curb chaotic city development." At a Beijing urban planning conference, The Construction Minister warned that "continued unplanned development could have disastrous consequences on Chinese urban development."

In addition to dynamic economic growth rates and a 20% increase in urbanization in 1992, China is also making sweeping historic changes in its once monolithic centrally planned economy. Two changes worth noting are: 1) the establishment of a single currency system, and 2) the opening of China's mineral resources to foreign exploration and extraction.

The use of FECs, Foreign Exchange Certificates, has been abolished, and the foreign exchange control system is being completely overhauled to permit easier exchange between the Chinese Renminbi Yuan and foreign currency. Previous swap markets, that have been at best problematic, are being replaced this month by a new system of seven experimental inter-bank markets throughout the country. (China Daily 2/19/94)

Perhaps of greatest interest to the Committee on Energy and Natural Resources is the end of the National Government's centrally planned and managed recovery of mineral resources. According to the Ministry of Geology and Mineral Resources, "Whoever offers the lowest input and best economic and social benefits will be allowed to develop the mineral resources in a specific area." (China Daily, 2/1/94) This new policy may signal the beginning of the GREATEST "Gold Rush" in human history. As if opening the mineral resources of China to foreign mining and extraction companies was not enough, the Foreign Affairs Office of China's State Council has announced that "foreign investors will be allowed to operate in more economic sectors in China from which they were once barred . . . (this includes) power plants, railways, airports, expressways . . ." (China Daily 1/31/94)

IV. HOW THE CHANGE IS IMPACTING ENVIRONMENTAL MANAGEMENT SYSTEMS

Economic growth is occurring so fast that environmental planning and infrastructure are unable to keep pace. Not only are cities being transformed, but patterns of agriculture, thousands of years old, are being overwhelmed by market forces as rural industries randomly sprout up at alarming rates. In 1993, China's export orient firms in rural industries "saw sharp growth of 60 per cent over 1992 . . . total output and profits of rural enterprises both grew by at least 40 per cent last year." For a rural population of nearly one billion people, the implications are boggling. As a result, the National EPA is trying to move trained environmental "extension agents" into "the fields" as quickly as possible to meet this unprecedented explosion of activity.

One example of how growth outstrips urban environmental infrastructure is in Haikou, the capital of Hainan, an island province, that is a newly created special economic zone. Haikou officials are receiving several development applications a day, and "as a result, demand for water and electricity doubled in a year (1992), creating daily power and water cuts." (CD, 2/15/94).

In response to this situation, leading Chinese environmental figures are sounding a call for action. Former NEPA Administrator and current Chairman of Environmental Protection Committee of National People's Congress is urging China to increase GNP spending on environment from its current level of 0.85% to 1.5%. Yet even 1.5% will most likely fall short of needed infrastructure as cities and industries continue to grow exponentially.

V. EFFECTS ON AIR & WATER AND THE IMPACT ON PUBLIC HEALTH

While China continues its second decade of double digit economic growth rates, the effects of this activity are being seen in both air and water quality, with grave consequences on public health. "According to World Bank reports, chronic obstructive pulmonary disease caused by exposure to fine suspended particles and sulfur dioxide was the leading cause of death in China in 1988." (The Nation, 2/14/94). Certainly this trend has continued over the last six years as evidenced in Shenyang where mortality rates for respiratory diseases increased 255% from 1983 to 1990.

Below is an excerpt on "Environmental Pollution and Human Health" from NEPA's "Report on the State of the Environment in China—1992."

7. Environmental pollution and human health

The mortality rate of the country in 1992 was 664 per 100,000 people, 0.9 percent lower than the previous year. Cancer is the primary cause of death in cities. The cancer death rate was 125.76 per 100,000 people, or 21.66 percent of the total death rate. The cancer mortality rate in big cities was 132.55 per 100,000 people, 99.48 per 100,000 people in small and medium-sized cities. Lung cancer caused the greatest number of deaths, which was 33.64 per 100,000 people, with 36.42 per 100,000 people in big cities, and 22.88 per 100,000 people in small and medium-sized cities. Compared with 1988, the cancer mortality rate in urban areas rose by 5.6 percent. Lung cancer death rates rose by 16.6 percent. In rural areas, cancer mortality rates have kept rising every year, from 95.02 per 100,000 people in 1988 to 102.53 per 100,000 people in 1992, an increase to 16.2 percent in the percentage of the total mortality rate. It has become the Number 3 cause of death in rural areas. Respiratory diseases were the Number 1 cause of death of rural residents. Deaths from respiratory diseases in 1992 were 168.32 per 100,000 people, accounting for 26.6 percent of the total number of deaths. Environmental pollution and unhealthy lifestyle and habits were the main causes of diseases.

In the field of water quality, though the attributable mortality rates can not compare to respiratory disease, environmental pollution is equally serious. The following information has been abstracted from an Industry Sector Analysis on Municipal Waste Water Treatment prepared by the Foreign Commercial Service, Beijing Office.

While problems in water quality exist throughout China, northern and northeastern China also experience water shortages which pose difficult challenges to public health, and in the near future, shortages may disrupt agriculture and industry. More specifically, industrial growth and urban population growth have caused water shortages in nearly 300 of China's 517 cities. Lack of adequate water treatment both for waste water and domestic potable supply need urgent action. According to Ministry of Public Health only one in seven rural Chinese have safe drinking water—which means more than 800 million Chinese rural residents (more than three times the U.S. population) have unsafe drinking water.

In the field of solid waste, environmental problems affecting the land, air, and water are deteriorating. The following information has been abstracted from a draft copy of an Industry Sector Analysis on Solid Waste Management prepared by the Foreign Commercial Service, Beijing Office.

In 1992, when China recorded a 25% jump in industrial growth, the country produced more than 587 million tons of industrial waste and close to 100 million tons of municipal waste. Of this volume of waste, only 40% of industrial and 3% of municipal was treated or recycled. Most visitors to China are often struck by the amount of untreated accumulated waste strewn about roadsides, villages, and fields. Unregulated landfills in the countryside often compound problems by contaminating farm land. As the volume of accumulated waste continues to grow larger, China's land, air, and water are increasingly endangered.

Mr. Chairman the above information presents a mosaic of the complexity and magnitude of change presently transforming China. Despite the hard efforts of many agencies in China, environmental conditions worsen day by day as a result of continued industrialization and urbanization. The United States can support the Chinese people in their efforts to obtain a better quality of life by helping protect China's environmental quality. Your call to end restrictions that prohibit the U.S. Asia Environmental Partnership, the Trade Development Agency, the Overseas Private Investment Corporation from operating in China is most welcomed. We hope the Administration and other Members of Congress will follow your lead and permit all U.S. programs to actively work in China for the better of our global environment.

This Hearing today is an important step to reforming our misguided environmental policy with China.

To conclude Mr. Chairman, I understand you recently led a delegation of U.S. Senators to China, and I wish to thank you for actively participating in China's economic transformation. Additionally, I thank you for this opportunity to provide testimony to your Committee. In the future the China Environmental Fund is prepared to support your work by keeping Members and staff of the Committee informed of activities in China, and as appropriate, helping coordinate future itineraries in China. Please contact us anytime to assist your efforts to promote economic growth and enhance environmental quality in China.

Wisconsin Office: P.O. Box 345, Mount Horeb, WI 53572, Tel. 608-767-3888, Fax 608-767-3887.

California Office: P.O. Box 5300, N. Hollywood, CA 91616, Tel. 818-761-3888, Fax 818-763-3790.

Mr. ELDRED. Apropos, Senator, of your earlier question as to whether the United States was asleep at the switch or not in its involvement in Asia and in China, I would offer the following statistic. According to a World Bank document the Japanese Overseas Economic Cooperation Fund is spending approximately $651 million on environmental issues alone in Asia. The figure listed for the United States is $62 million. That may not include funds from the Import-Export Bank, but it is clearly a 10 to 1 ratio. The Japanese are outloaning and granting the United States in Asia by a margin of 10 to 1.

I would like to talk a little bit about some of the environmental conditions that we have observed in Shenyang, China, and they are appalling. They stagger the imagination.

Today in Shenyang alone there are more motor vehicles than there were in all of China in 1979. The 200,000 vehicles now on the road cause massive traffic jams, and few of them are equipped with even the most rudimentary emission control devices. By the year 2000, Shenyang will have as many as 1 million vehicles.

Drinking water—Shenyang's drinking water is not safe. 91 percent of wells sampled failed to meet national standards for arsenic, cyanide, ammonia, nitrogen, phenols, and oils. Intensive use of groundwater in Shenyang has caused the underlying water table to drop 33 meters, and this decline is proceeding at an annual rate of 1.5 meters per year.

Many drinking water wells are located in or near industrial areas, or draw on groundwater recharged from the highly polluted Hun River into which untreated domestic sewage and industrial effluent flows. There is currently no wastewater treatment, although one plant is under construction that is being built by the French.

In the area of air quality, according to the World Bank, Shenyang has the world's highest concentration of soot and sulfur dioxide pollution. Although all parts of the city have exceeded standards for sulfur dioxide, industrial districts have reported concentrations three to six times higher than U.S. national ambient air quality standards.

The concentration of total suspended particulates was 9 to 13 times greater than the U.S. standard. Now, about 70 percent of these particles of soot are small enough to lodge permanently in human lung tissue, and are the ones that are regarded by health officials as being the most pernicious.

The CHAIRMAN. How does this compare with Los Angeles on a bad day.

Mr. ELDRED. Marc lives in Los Angeles, so I think I will let him comment on that.

Mr. BRODY. It really does not, sir. Just on a personal note, when I met my friend at the airport he had been there 1 week, and he was already sniffling, carrying tissues with him. We have not experienced air pollution like this in the United States.

The particulate matter in the air is the leading cause of death in China. I might add that according to World Bank reports, in 1988 fine suspended particles and sulfur dioxide was a leading cause of death in China. And when you look at Shenyang since 1988 I can provide these statistics—that Shenyang's mortality rates for respiratory diseases increased 255 percent from 1990 to 1993.

So, we have an expression. It is darkness at noon in Shenyang right now. It literally is, and World Bank reports have shown that Shenyang and Beijing absolutely disappear from sight from satellite photography at certain times because of the dense particulate pollution in the air.

Mr. ELDRED. Let me just go on for a moment about energy, and then I will be finished.

The operation of the Shenyang economic engine results in the combustion of 1.4 tons of coal for every man, woman, and child in the city. This compares to about just under 1 ton in the Nation as a whole.

Shenyang used 13 million tons of coal equivalent during 1993 to fuel the industrial, domestic, and transportation sectors of the economy. Of this, 9.7 million tons was actually coal.

This will grow to 14.5 million tons of coal by the year 2000 if energy use continues to grow at the current annual rate of 5 to 6 percent, and coal maintains its proportional contribution to the mix of fuel sources now used. An analysis of national energy trend data indicates that this is likely to be true.

I think that there has been a lot of discussion in the last 2 years about world environmental conditions and greenhouse gases. What is missing from that discussion is the realization that only one-twentieth of the capital resources that are necessary to seriously address this problem are in place by world financial institutions.

And to me that is a very significant fact. It means that 95 percent of the projects that are needed are not just unfunded but it does not seem like anyone in particular is working on developing the international institutions that are going to be necessary to make that funding available. In other words, the question is not being asked. And the conclusion on our part is that this has got to come from the private sector.

Mr. BRODY. I would like to add one comment on that. The change in China is so profound, this industrialization, that we have likened it, and others have likened it, to essentially the rebuilding of Europe. The magnitude of change that is taking place in China is like the rebuilding of Europe after World War II. And perhaps under the leadership of the United States, we can embark a new environmental Marshall Plan, a green Marshall Plan, not just for the benefit of one-fifth of humanity but essentially for the benefit of our global environment.

And I think, again, going back to your very good point, I hope your comments are a watershed for us in American policy to now come to grips with the inconsistencies we have and to turn our attention to helping people—not just helping national governments but helping people by permitting very fine organizations in the United States and agencies to operate in China to address pressing needs.

And I would very much ask for your support, the committee's support, to develop some type of Marshall Plan for China, some type of environmental plan, green plan for China that will address the shortfall of capital and private sector involvement that Mr. Eldred has alluded to.

The CHAIRMAN. I wish everybody in the Government could hear your testimony.

Mr. Chandler.

STATEMENT OF WILLIAM U. CHANDLER, DIRECTOR, ADVANCED INTERNATIONAL STUDIES UNIT, BATTELLE, PACIFIC NORTHWEST LABORATORIES

Mr. CHANDLER. Mr. Chairman, thank you for having me back. I agree that it is difficult to overstate the environmental impact of China's energy use, both locally and globally. Locally, I think you have seen it. You have been in cities like Taiyuan, and the Shanxi Province, and Shenyang, and you have seen the particulate levels that are 20 times the maximum allowed in the United States.

And you know the impact of lung disease on people in China, which is four times what you would expect in the West. Or the acid rain levels in the cities, which give you a pH of rainfall of about 4, which is about the level of tomato juice in acidity.

And globally, in terms of carbon dioxide emissions, I think the real question there is not whether China will increase its CO2 emissions, but whether they will triple them or double them over the next quarter century. And to the extent that the United States can help China achieve that lower level, we will help ourselves.

China does have options, energy efficiency being one, that they have taken seriously over the last 15 years. They have done something that no other developing country at its stage of development has done, which is to cut the rate of energy demand growth to only half that of the economy. The question is, will they have the technology and the financing to continue that progress?

Natural gas is a different story, I think. The central planners have not deemed fit to allocate much in the way of investment to exploring and developing that resource, and that is a crying need, I believe.

In renewables, there is a big opportunity in the 8 quadrillion Btu of firewood and crop waste the rural Chinese consume, an opportunity in using our technology to convert that much more efficiency and with much less environmental impact.

Let me talk very quickly about two or three things that I would like to see us do to improve the situation. One relates to the Beijing Energy Conservation Center, and I suppose I should start by saying thank you for the U.S. Government's support for that operation.

A year ago, almost exactly a year ago, I testified here and I mentioned that we would like to try to create this center to do policy analysis and joint ventures development in energy efficiency in China, and with the support of the Department of Energy and the Environmental Protection Agency and the World Wildlife Fund in December we did open this operation which will be run by and for Chinese to help promote the energy efficiency business in China.

One of the things I hope they can do is to perform like some of the centers we have created in Eastern Europe, which have more than paid for the tax revenues that went into them by helping arrange contracts between American energy efficiency companies and local clients. The center in Prague has developed more than $30 million worth of business for a single American company.

Creating the market for energy efficiency is something that this center and U.S. Government can help local governments and utilities do in China through transferring our policies, things that we have innovated like integrated resources planning which our utilities use to put efficiency on an equal footing with supply, to provide incentives so that companies like Owens-Corning and sell the insulation that it will manufacture there, the controls that Honeywell and Johnston Controls will be manufacturing there, and the variety of projects that range from Armstrong steam traps to other electric motors. Creating the market so that they can be successful in their marketing efforts is a big priority for us.

And the last thing I would like to mention is encouraging the opening of the natural gas sector in China to U.S. participation. U.S. companies have the expertise and the financing to help, No. 1, resolve the question of how much natural gas is really there and, No. 2, to develop that clean burning, low carbon emission resource to the extent possible.

Finally, I would like to say that a lot of the information that I have included in my written testimony was derived from a recent paper prepared by Jessica Hamburger at Battelle, a paper called "Linking Energy and Environmental Policy in China," and I would like to offer that to you for attachment to the record if you would like.

[The prepared statement of Mr. Chandler follows:]

PREPARED STATEMENT OF WILLIAM U. CHANDLER, DIRECTOR, ADVANCED INTERNATIONAL STUDIES UNIT, BATTELLE, PACIFIC NORTHWEST LABORATORIES

The Chinese economy is growing explosively, expanding an average 9 percent annually since economic reforms were initiated in 1978. Coal has fueled that growth. Efficiency improvements have helped hold the surge in energy use to only half that of the economy. Nevertheless, annual coal use in China has almost doubled over the last fifteen years, and is likely to double again by 2010.

China's rapid energy demand growth threatens severe environmental damage. Coal, the primary fuel in China, is used inefficiently and without pollution controls, causing serious human health problems, damaging crops and buildings, and raising the risk of global climate change. Chinese use less than one-twelfth as much energy per capita as Americans. Yet, China now ranks second in the world in energy-related greenhouse gas emissions—behind only the United States—and will become the world's largest emitter by 2025. Global environmental problems cannot be redressed effectively without China's participation.

China also ranks among the least energy-efficient nations in the world. China's experiment with central planning left a legacy of energy waste and energy shortage, but its shift to market mechanisms offers the prospect that both economic and environmental needs can be met more effectively.

Bilateral cooperation with China could help effect a number of positive results. Support for development and utilization of U.S. policies such as Integrated Resources Planning could help create the market for demand-side technologies ranging from electronic controls to steam traps to insulation. U.S. industry is beginning to supply China with energy efficiency equipment, ranging from insulation to valves and controls. A new, relatively independent institution, the Beijing Energy Efficiency Center, could serve as a forum for technology transfer and market development that could help minimize the economic and environmental impact of China's exponential energy demand growth.

RAPID ENERGY DEMAND GROWTH

Coal supplies three-quarters of China's energy. It has fueled the nation's explosive economic growth, which has averaged 9 percent annually since the economic reforms were initiated in 1978. That means China's economy has grown by 250 percent in the last 15 years. Astonishingly, China's energy use has grown "only" half as fast. No other developing country has ever achieved such a feat.

Still, China's energy demand growth is prodigious. Coal production almost doubled from 620 million tons in 1978 to more than 1.1 billion tons last year. That increase is comparable to total coal production in Russia.

Chinese oil and gas use remains very low by comparison. The country uses oil at a rate of just over 2.2 million barrels per day (4.75 quadrillion BTU), and gas at a rate of less than 300,000 barrels oil equivalent per day (0.6 quadrillion BTU).

LOW LIVING STANDARDS

China uses one-twelfth as much energy per capita as the United States. Because most of China's energy is used in industry, Chinese households use only three percent as much energy as American ones. Central heating is often unavailable, and air conditioning rarely exists outside of hotels.

But with rapid income growth, many Chinese are buying appliances. They purchased almost 7 million refrigerators in 1992, one-fifth more than in 1991. The residential share of electric power consumption tripled to 8 percent during the eighties. The implication is clear: rapid economic growth will only increase energy use, and in turn the environmental burden of energy production and conversion.

ENVIRONMENTAL STRESS

Heavy reliance on coal produces a litany of environmental risks for China, the region, and even the globe. Local pollution problems long ago reached crisis proportions.

> Most major Chinese cities exceed international air quality standards for particulates due to coal combustion emissions. Cities like Taiyuan in the Shanxi Province experience average wintertime particulate concentrations 20 times greater than the U.S. maximum daily standard.
>
> Chronic obstructive pulmonary disease (COPD) in Shenyang is four times higher than in developed countries. Age-adjusted mortality from lung cancer is among the highest in the world. Air pollution is estimated to account for three-fifths of COPD and one-quarter of lung cancer.
>
> Precipitation pH is around 4 in most cities—the acidity of tomato juice. Acid rain in south China reduces the useful life of zinc-plated steel to only one-seventh of normal.
>
> Nitrogen oxide concentrations in Beijing, Tianjin, Guangzhou, and other major Chinese cities average almost 20 micrograms per cubic meter, compared to micrograms per cubic meter in Los Angeles.
>
> The countryside has been deforested in part by heavy use of firewood for household fuel. Biomass supplies four-fifths of China's rural household's energy—almost 8 quadrillion BTU of firewood and crop stalks in 1990.

China's carbon dioxide emissions pose significant risks for climatic change. That is, China's energy future can directly affect America's environmental future.

Coal produces twice as much carbon dioxide per unit of energy as natural gas. China now ranks second in global energy-related carbon emissions. The basic climate policy question for China is not whether its carbon dioxide emissions will increase, but how much they will increase. Plausible projections suggest that China could triple its carbon emissions in just 25 years. With work, that figure could be held to a serious, but more manageable, doubling.

Several cubic meters of methane are released for every ton of coal produced, that is, several billion cubic meters in 1992. Only a small portion of coal bed methane is used as fuel, the rest escaping into the atmosphere and contributing to the greenhouse effect. U.S. assistance could help ameliorate this and each of the above problems.

POLICY AND COOPERATION

China's options for reducing the environmental impacts of rapid energy demand growth include energy efficiency, switching from coal to renewables, natural gas, and improved coal combustion technology.

Chinese energy policy is moderately helpful on these issues. Subsidies to the coal industry are to be phased out. Oil and electricity prices have been increased toward market levels, though not always as quickly as western economists might have liked. Most importantly, the Chinese economy is being transformed into one that resembles a real market: Almost half of all production comes from the non-state sector. At the same time, the Chinese government limits development of natural gas both by allocating small sums to, and restricting the participation of, foreign firms in gas resource exploration and development.

No U.S. policy is more important for continued reform of the Chinese energy sector than continued opening of China's economy to trade and market competition. Continuing market reform will inevitably mean liberalization of China's political system, fewer greenhouse gas emissions than we will have otherwise, and major market opportunities for American firms.

The U.S. government could take additional steps to help create the market for energy efficiency and clean fuels in China. We can start by providing modest resources to support relatively independent energy efficiency programs. A major opportunity resides in supporting the Beijing Energy Efficiency Center (BECON), a nonprofit, non-governmental organization designed to promote energy efficiency to achieve the twin goals of economic development and environmental protection.

BECON was founded by the U.S. Department of Energy's Pacific Northwest Laboratory and Lawrence Berkeley Laboratory, the World Wildlife Fund, and the Energy Research Institute of the State Planning Commission of China. The center builds on a strong analytical base to promote energy efficiency in four areas: policy research and development, private business development, training and demonstration projects, and public education. It is directed and staffed by Chinese energy experts and provides assistance to foreign firms wishing to tap into China's energy efficiency market.

BECON could, with modest additional resources, help create the market for American products including electronic control devices, insulation, steam traps, electric motors, efficient boilers, windmills, coal-bed methane recovery, biomass regeneration and gasification, as well as the transfer of American-style energy management systems such as Integrated Resources Planning (IRP). Our utilities innovated IRP to help put demand side management on an equal footing with supply acquisitions. Few options could be more important to the future of Chinese utilities than the adoption of IRP methods.

Supporting BECON's energy efficiency efforts could help reduce global carbon emissions, minimize local pollution and increase the market for U.S. exports. And because BECON is intended to become an independent, non-governmental organization, its success would foster democratization.

Opening China's energy market to U.S. involvement in the discovery and development of Chinese natural gas resources could have major economic and environmental benefits. China lacks the capital and the expertise to develop its gas, and U.S. firms could help provide both. Natural gas is the cleanest burning fossil fuel, and among them produces the least carbon dioxide. The U.S. government could encourage the Chinese government to allow U.S. companies to participate in the gas business in China as part of our efforts to establish freer trade.

I would like to thank the Senate Energy Committee for this opportunity to present my views.

China: Primary Energy Use by Source
1952-1990

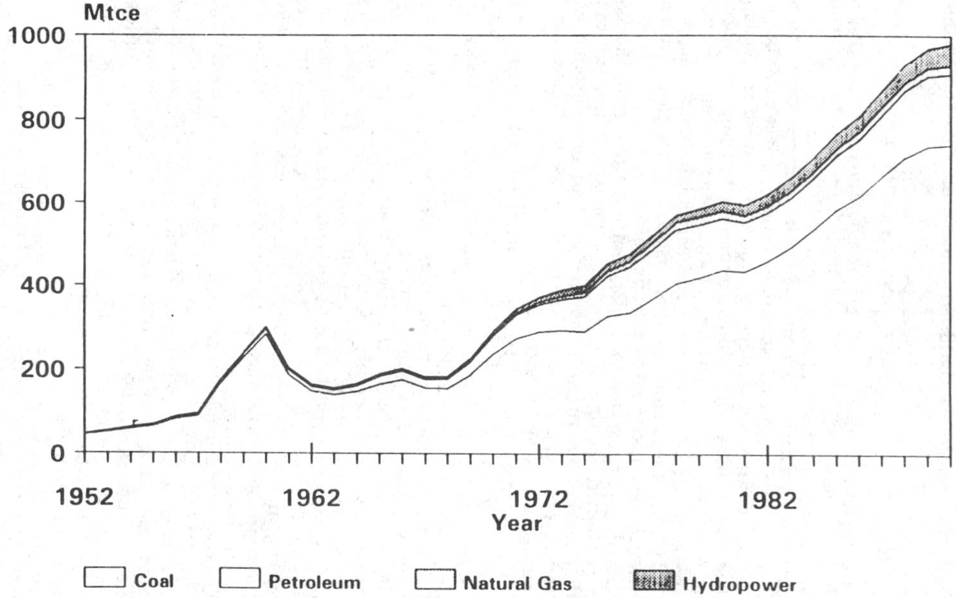

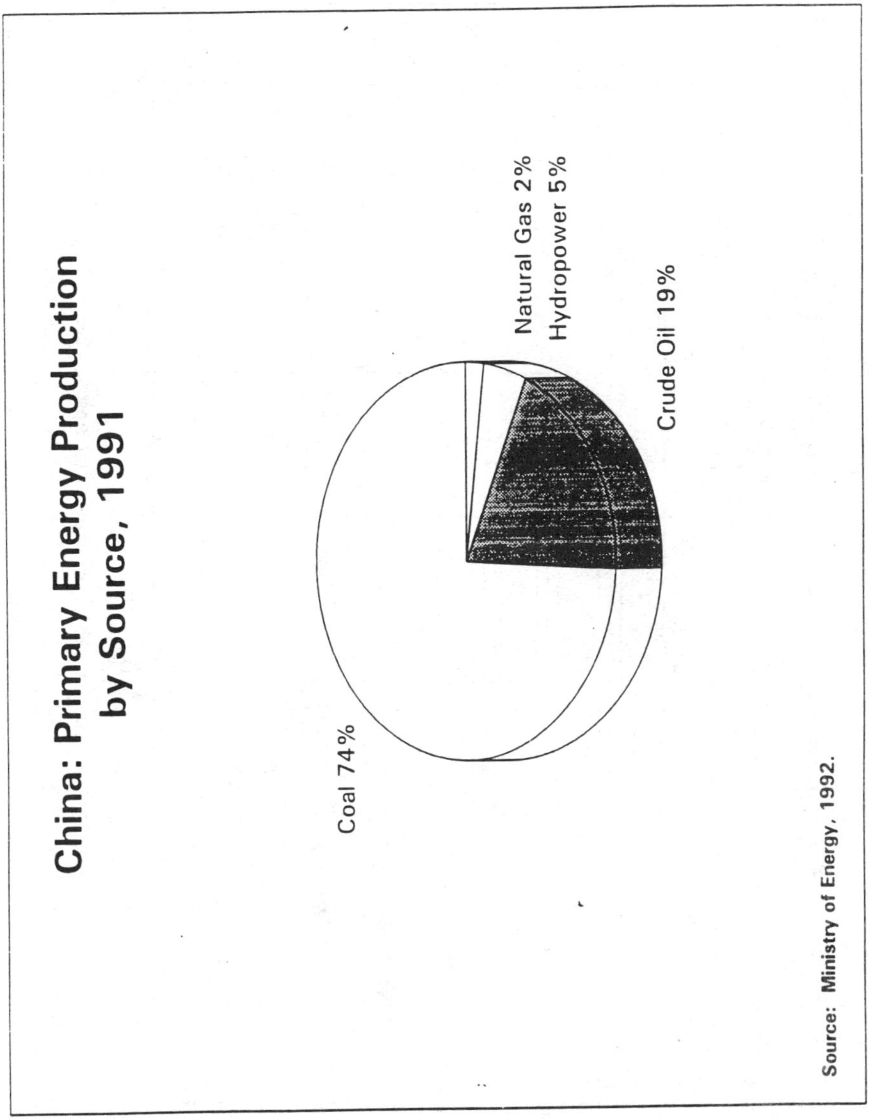
China: Primary Energy Production
by Source, 1991
Natural Gas 2%
Hydropower 5%
Crude Oil 19%
Coal 74%
Source: Ministry of Energy, 1992.

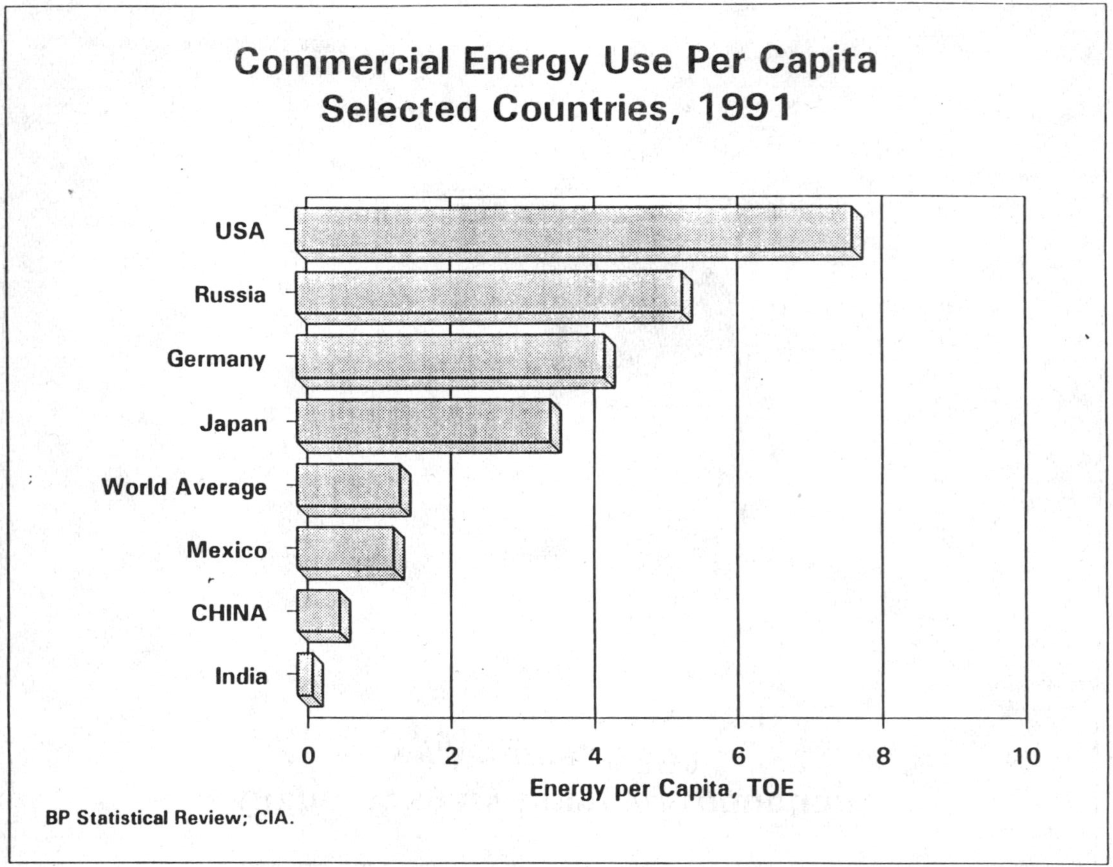
Commercial Energy Use Per Capita
Selected Countries, 1991
USA
Russia
Germany
Japan
World Average
Mexico
CHINA
India
0
2
4
6
8
10
Energy per Capita, TOE
BP Statistical Review; CIA.

Shares of Global Carbon Emissions, 1991
(Percent)

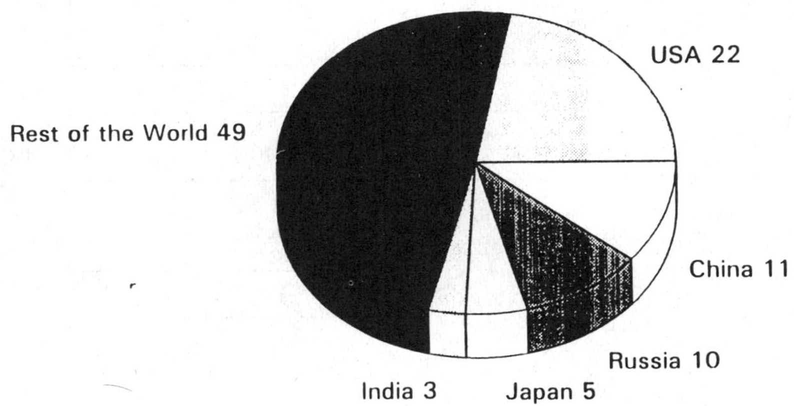

Source: CDIAC, ORNL; PNL

China: C02 Emissions Projections
Base Case and Efficiency Scenarios

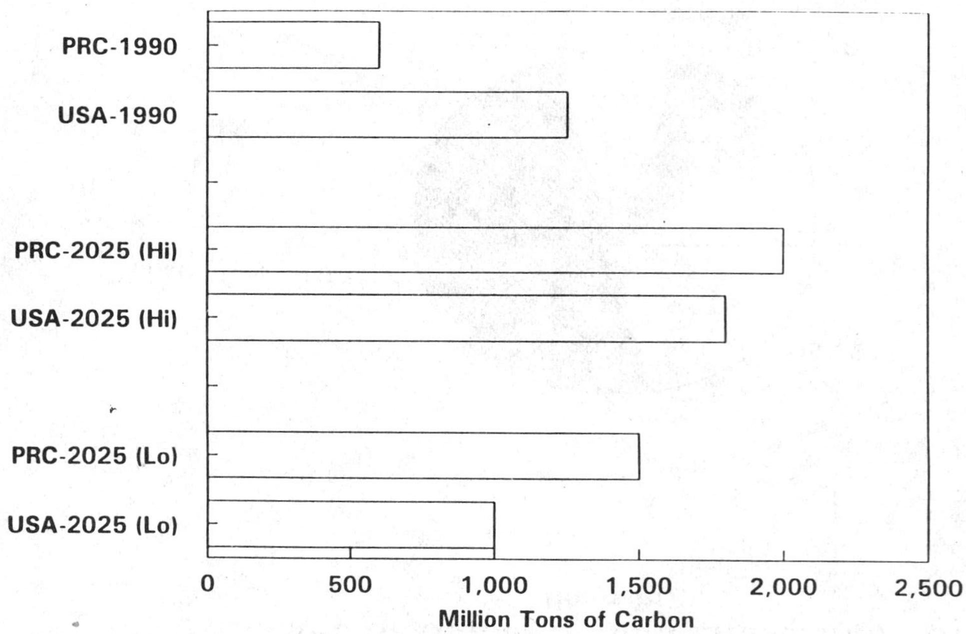

Source: LBL; W. Chandler.

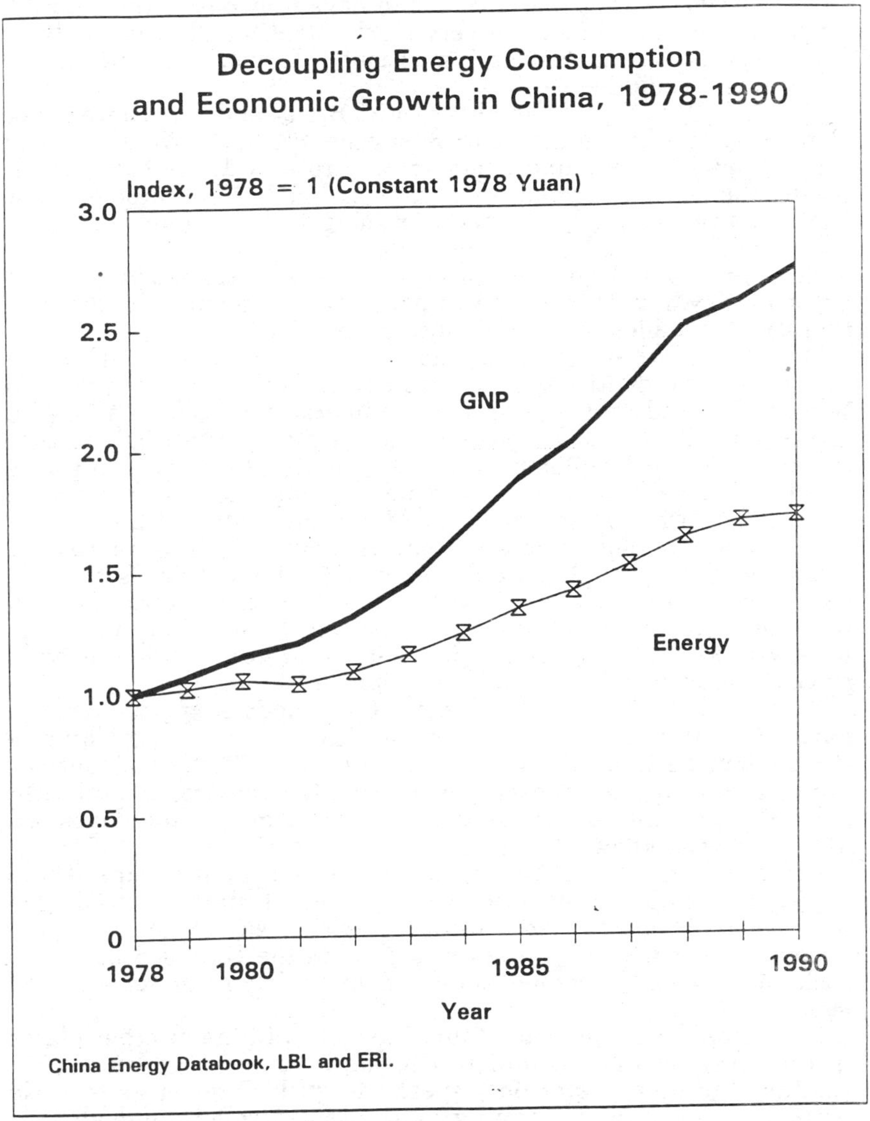
Decoupling Energy Consumption
and Economic Growth in China, 1978-1990
Index, 1978 = 1 (Constant 1978 Yuan)
3.0
2.5
2.0
1.5
1.0
0.5
0
GNP
Energy
1978
1980
1985
1990
Year
China Energy Databook, LBL and ERI.

The CHAIRMAN. Yes, I would like to have that put in the record.*

The CHAIRMAN. Thank you very much, Mr. Chandler. It has been very, very powerful. And as I mentioned, I wish everyone in the Government could hear it.

Now, Mr. Wing, you have heard Mr. Brody and Mr. Eldred talk about this terrible situation in Shenyang Province. We also know that China last year had a net growth rate of 13 percent, which means that they are going to continue to add to the problem, not subtract from it. At least they are going to be adding additional energy and projects.

Now, how would you, as an entrepreneur, accommodate to the problem they have? How can you participate in solving the problem that they have identified in Shenyang Province?

Mr. WING. What we would do, Senator, is to work with the government to identify the best system, the best electric system to serve their needs both on an environmental and a cost basis. You are very familiar with the project that we did over in England that got the U.K. in compliance with EC quality standards, and they could not have gotten into compliance going the way they were going to go with coal. It simply would not have gotten done.

So, first we would identify what is doable. If it were coal, it would be clean coal. We would build it to the standards that we build here. We certainly would shake out the alternative, on whether we could build gas plants. As you know, along the coast we are looking at several gas fired plants. We are looking at wind power. There are a lot of good wind sites in China.

To get from the conceptual stage to the do-it stage, we have to have the ability to make the technological transfers into China of the equipment that will solve these problems. Those technologies are strongest in this country, yet the supply of environmental technologies from this country is the smallest of any of the developed nations, as you know.

I mean, there is no question that we can fix the problems. There is just no question. I mean, look at the United States. You alluded to it earlier, Senator. We continue to pick away at the marginal amount of—be it particulates or sulphur in the United States, and each time we are spending more and more dollars to get less and less.

If we took the same amount and spent it in China or other places in the Third World, assuming—I mean, no one has convinced me yet that the ozone layer only speaks English. I see it as a world problem. I see global warming as a world problem, and all that money could be spent a hell of a lot better where you are getting a bigger bite of the apple to get that out.

Why that is not good public policy in the United States is something that I just simply cannot understand. We know how to do it, and we can get it done.

The CHAIRMAN. Well, Mr. Eldred and Mr. Brody called for a Marshall Plan for environment in China. I think that is very appropriate in terms of the effort, but in terms of the foreign aid, a country that is growing at the rate of 13 percent last year really does

*See appendix II.

not need foreign aid, as Europe did to rebuild. They need different kinds of help.

And what I am saying is, how do we adjust American policy to get this done?

Mr. WING. They can self-finance. I mean, the projects can carry themselves. As they privatize and they need the power, you supply the power and clean up the air and the water at the same time. I mean, why we are doing good in this industry and in this country is that we bring the technology and the ability to put it together, and we bring the money with us.

Mr. BRODY. If I may, Mr. Chairman, I would like to follow up on Mr. Wing's remarks, and especially your last comment. I do not think this has be any type of foreign assistance or foreign aid. As you suggest, China's economy is growing so rapidly that this is something that they can just build into their system.

The efficiencies that American technology can offer the Chinese will save them money, they will not cost them money. They will save them money almost from day-one. We can start using certain types of efficiencies. And we have talked with the people in Shenyang about fuel allocation using different types of coal, different types of fuels depending on the seasons.

It is known that there are times of the year when the inversion layer is so bad in Shenyang that certain types of policies can make profound influences on environmental conditions.

I would just like to suggest to Mr. Wing that we would like to invite him to Shenyang, and all others. We are working with the municipality of Shenyang. We have with us now here in Washington the vice mayor of Shenyang and the director of the Shenyang Environmental Protection Bureau. We are holding a briefing that Mr. Wing and anyone else is invited to tomorrow, where Shenyang will be present, the municipal officials, to review their conditions.

And with your permission, Mr. Chairman, I would like to submit a copy of our environmental assessment on Shenyang that speaks to the specific conditions there, but also speaks to the broader questions for all of China.*

The CHAIRMAN. Well, you mentioned that Japan put up $651 million in loans and grants in environmental matters in Asia, while we put up $6.2 million. That is part of the problem. So, we need, what, Exim Bank, and we need TDA, and we need OPIC. Those are three things. What else do we need to do to get involved in this problem?

Mr. WING. Without question we need to get the Most Favored Nations thing behind us and on a permanent basis. We simply cannot work on projects that are 40 and 50 years long, that are multibillion dollar commitments for those of us that are in the United States, wondering whether the ladder is going to be pulled out from under us every June 3. It simply will not work. We simply will not have the jobs and the economic growth here, and we will go somewhere else.

I would hope that logic would follow, sir, that if we get permanent Most Favored Nations, and we do not use this blunt instrument anymore, that the others would follow.

---

* Retained in committee files.

Clearly, Exim has to deal with project risk in China. Clearly we need OPIC. We do not need it as much for the overall, absolute amount of dollar commitment, but to show that the U.S. Government and that this country in total is behind this effort. That is my estimate of it.

And we are right on the cusp of this. It is going to be the biggest opportunity we face, and the people—unfortunately at this point I believe we have done everything that we can hopefully right, but the decisions that are made here are either going to make it all work or it is all going to be just a dismal failure.

The CHAIRMAN. Now, should we not have joint implementation? Are you all familiar with joint implementation? That is where industry would use its trade-offs, if the Government would allow it, to spend for example in China rather than the United States. Rather than spending that last billion dollars to clean up that one-tenth of 1 percent of the particulate matter, you use it to clean up the first 95 percent in China.

Mr. ELDRED. Any economist, Senator, would agree that a dollar spent in Asia—you get much more bang for the buck in Asia than you do here in the United States.

If I may just back up a moment to your earlier question about how to get this done, one way to get this done internationally is to look at the history of how this was done in the United States. During the sixties and seventies when wastewater treatment plants were built in the United States in large volume, they were built through the municipal bond market, essentially using revenue bonds where the user fees would secure payments to the bonds and would allow the accumulation of large amounts of capital for a single expenditure.

Now that institutional framework does not really exist in an international market that is simple and easy to implement. And that is one source of capital that I think needs to be worked on and developed.

That is a long-term strategy, something that would require a great deal of research, but it is a way of taking investment money from the Western countries and funneling it to the developing nations in the Pacific Rim using a fairly accepted set of procedures in the West. And that would raise an enormous amount of funds.

Another option that is being used by some utilities now, utilities in the United States, is carbon sequestration, which involves payments made by utilities for the production of biomass in the southern hemisphere, and that raises capital. It is more difficult to apply in China because the weather in Northern China in particular is not quite so favorable. But that is yet another source of capital.

The CHAIRMAN. One of the problems, I think you identified it, Mr. Brody, is that nobody really is in charge in this country of the environment in China, the environmental challenge in China. I mean, we have got DOE that does a little bit with respect to technologies, and you have got EPA that monitors the problem, and you have got Commerce that talks about sales of goods.

But nobody attacks the problem of the environment in China as a coherent, holistic problem, and not only not as a Marshall Plan, but even as a problem that people are aware of. And it seems to me that we need to look at the environment in China as a separate problem, even separate from MFN. I hope, Mr. Wing, we can get beyond this annual MFN, shooting ourselves in our own foot, and shooting the polluted Chinese underclass in the lung in effect, by depriving them of a cleaner environment.

But when we get beyond that, then we ought to get a coordinated plan to apply international and national resources to this problem, where there is such a tremendous ability to make progress.

Mr. CHANDLER. Mr. Chairman, even a modest increase in resources to work with China would be helpful, especially if you use that money to work with local experts. If you work with local experts, you can get 10 times as much per dollar as if you try to do it all unilaterally. If you get local experts to work with you, then you can work on those policy issues, creating the infrastructure that you have to have in place before energy efficiency companies can do business.

Mr. BRODY. I would like to return to your comments. I think there are two opportunities or even two current efforts undertaken by the U.S. Government that can try to synthesize or coordinate a broader U.S. policy approach to China's environment.

The first is President Clinton and the Department of Commerce, DOE, and EPA have created an environmental technologies working group to promote U.S. exports in this field, and I am certain that the process of this interagency working group will provide the means to address these types of issues. So, I would support the work of that interagency working group.

Secondly, there is one organization that could potentially serve as an umbrella which is the U.S.-Asia Environmental Partnership. Though it is quite new, the U.S.-Asia Environmental Partnership does have a broad brush approach. It has many types of programs that look at environmental issues in other parts of Asia but, again, unfortunately is presently barred from China.

The CHAIRMAN. It is barred from China?

Mr. BRODY. It is barred from China. The U.S.-Asia environmental partnership, with the word Asia, is barred from the most significant place in Asia. And it is, again, one of those types of ironies, inconsistencies, fundamental problems we have.

The CHAIRMAN. That is related to the MFN?

Mr. BRODY. No, it is a separate issue. It is a Congressional statute right now. President Clinton I believe has Presidential authority to waiver that restriction, but it has not been exercised. But it really is a misallocation of resources. I mean, balls are coming across the plate and we are not even swinging in. We are going out on called strikes, we are even swinging.

And what I would like to add is that if things were not bad enough in China's environment today, in my written testimony I cited that for the first time just a month ago the National Government of China has opened up its entire mineral resources to the foreign world. That has never happened in the modern state of China. It has always been a centrally planned economy, state-managed organizations.

Now they are refining programs, policies, procedures so that any company can get involved in exploration, extraction of mineral resources whether it is coal, oil, natural gas, any type of minerals, and they are just going tc essentially look at who gives them the best offer, who is going to have the best, in their minds, environmentally sensitive programs.

We are working with the national EPA, our organization to help establish both a compensation fee for resource extraction and some of the policies about resource extraction in China.

So, things are happening so quickly in China now that it is not just the growth, it is the social reform and the institutional reform, and the opening up of China to the outside world that I think this committee and all members of Congress need to be aware of. When they turn their mineral resources open to the world market, that is a very profound event, and I do not know how many people have taken notice of that.

The CHAIRMAN. The impression that I got was that the Chinese were open to environmental suggestions and technologies, and they want them. It is not that they are resistant, as the communists in Eastern Europe were during those days. They really want them. Do you share that view, Mr. Wing?

Mr. WING. Yes, sir, I do. I think there is an issue of how much they can afford how fast, but I think the realization is there. I mean, you have been in Beijing when you could not see from me to you, and they realize just how horrible it is going to get. And on that growth trajectory it will be—being twice as bad as it is now is about the best we can hope for.

So, they are there. I mean, if we get there with the right products and the right financial structuring it will get better. I mean, they will be with us.

One of the problems, and again we talked about it earlier, but if we can get it off of one Nation here and one Nation over there, what happens is as we attack the marginal amount here and we are trying to go from 12 parts per million of something or other to 9 parts per million, and we spend $5 million doing it, what happens is it makes that entity in the United States less and less competitive.

So, guess what happens? It moves offshore, where we are still at 400 parts per million, and we are all sitting here saying, who would have thunk it? We need a broader base. I do not know how to do it in the Government but you need a hero. You need one person where all of this comes together, and you are exactly right.

And if you can make one recommendation to the administration, somebody that understands this issue and the economics tied to it along with the energy parts, that can see—that does not expect to see cowboys and Indians when they go west of the Hudson River, that they are a little more international than that, that we can take a broader view. And I think we need that one person.

I do not believe the multiagency—too much falls through the cracks. It does not get done. We talk about it, and it does not get done.

To get it done, projects need to happen. It has to go from paper to actual pieces of hardware and changing of behavior. And that takes tough action and an individual that is—instead of us trying to drag the Government along behind us, that is pushing us along out in front, and it is achievable.

The CHAIRMAN. I have got to go make a speech at noon. So, I want to thank the panel very much. This has really been outstanding. I just wish everybody in the Government could hear this, including the President, because it is a solvable problem. It is a huge problem. It is a huge challenge for this country.

If we can just do what is obvious to be done, we can improve the world's environment more in this area than probably anything else that we can do. I mean, all this effort down at Rio, which is great, but this can be much more effective than that at less cost.

Mr. WING. It is solvable. It has to be solved.

The CHAIRMAN. I do not want to neglect the Rio initiatives. It is just that this can achieve so much more, much more quickly, and so much more obvious.

So, I want to thank all the panelists very much.

Mr. WING. You are welcome.

The CHAIRMAN. And, Senator Bennett, apologies to you for speaking so long.

STATEMENT OF HON. ROBERT F. BENNETT, U.S. SENATOR FROM UTAH

Senator BENNETT [presiding]. Not at all, Mr. Chairman. I endorse what you said. I wish everyone also could hear this. I have nothing profound to ask you. You have covered all of the issues that I have in mind. I simply want to go firmly on record as agreeing with my chairman that we have got to get the MFN thing behind us. We have got to make it permanent.

I have been sitting here all morning looking at an editorial in the New York Times entitled, "Standing Up to China." And the New York Times is not often my favorite newspaper, and it is not today.

I am reminded as I listen to this requirement, to this demand for us to stand up to China on some of these serious humanitarian issues, of the Wall Street Journal piece that started out quoting the old Mel Brooks movie, Blazing Saddles, where Mel Brooks opens the movie by placing a gun to his own head and then saying, "stop or I will shoot," and suggesting that that is a little bit of what happened to Secretary Christopher when he went to China. He rode in with the gun pointed at his own head.

The only comment I would make, and ask you to comment on briefly, is on the atmosphere in China as you understand it politically. If we should indeed cancel MFN would there be serious reprisals against the United States economically, and perhaps more importantly, would it change the Chinese mentality politically with respect to how we are viewed? Mr. Wing?

Mr. WING. I think without question it will impact us, first economically. If there is not Most Favored Nations, I would expect that in our area of business, in energy, that our share will be dramatically reduced; reduced only to those things which they believe they could only get from the United States, and I frankly do not know what that would be because of all the licensing agreements among our various companies, as Malcolm was saying, the GE licenses, the GEC and so on and so forth. So, I think it would have a horrible effect on our ability to be involved in the economic growth, particularly infrastructure.

I cannot see, anyway, if that happens why they would then turn around and say to us, but we would like to be your friend on strategic issues. I just do not think that that makes much sense.

I said to somebody earlier, Senator, that the last visit—if you were to imagine something in the United States equivalent to one of the two party's conventions plus and inauguration plus a state of the union address, that is kind of what the People's Congress is.

And if you switched it around and a foreign minister, not a head of state, insisted upon coming to the United States during such a process and insisted on taking serious issue publicly with everything you were doing. I think that you would suggest that that would not come to much good. In fact, you might question what it was you were trying to achieve. And I think that is precisely what has happened.

We need to understand that the gun sits in the middle of table, that we do not have it in our hand, we do not want to force them to put it in their hand, and to discuss these things as adults. I think if we cancel MFN we are heading down the road of a serious, serious problem economically. I do not see how we would leap over that to deal with strategic or environmental or human rights issues. I just do not see.

You are tugging on Superman's cape in terms of human reaction if you start out that direction.

Mr. CHANDLER. Senator Bennett, I have to confess that I like the New York Times, but I think that we would agree that the worst thing that could happen for reform in China, whether it is economic or political, is to undermine the people who are the strongest advocates of that reform, which is the people in the budding private sector.

Mr. BRODY. I would like to add that, talking about the budding private sector, I think that is an excellent comment. Cancelling MFN unquestionably would cause profound economic disruption in China. This is an economy that is moving from a state control to a market economy. They do not make changes that quickly.

America accounts for a huge share of Chinese commercial products or domestic products. All you have to do is go into a K-mart or a Wal-Mart and you will see how much China churns out. By cancelling MFN huge numbers of Chinese workers are going to wonder where their products are heading, and I do not know what will happen to factories in China.

So, not just for American business, but unquestionably Chinese factories will experience profound and sudden economic disruption. Certainly, that would cause some type of social repercussion.

The other thing I might add is it does send a signal that we are trying to dictate policy to China, which I think will be seriously, seriously taken as an insult and we will have repercussions.

I know that as a whole, Chinese people were terribly hurt when the U.S. Government did what it could to persuade the International Olympic Committee not to permit China to host the 2000 Olympics. I have worked with people in Beijing on the 2000 Olympic effort.

I think that an Olympics in Beijing or continued economic activity in China are the very things that will bring about the social reform that this administration is pursuing. And with all respect to the administration, I think, again, it is a contradictory policy. I think it is a dangerous policy. I would even say it could spiral into shifting a cold war approach that used to be the United States and the Soviet Union to the United States and China.

And the tragedy is the Chinese economy is so strong, growing so fast, has so much purchasing power that they offer perhaps one of the best if not the best hope and market opportunities to bring along the American economy. And there have been years when the American economy led the world economy. There is no question the Chinese economy will lead with world economy in the next century, and we ourselves can benefit on an individual basis here in America from that.

Cancelling MFN would be basically slapping our customers in the face. And I think it would be just encouraging the Chinese to work all the harder with what we would consider our new competition, the global environmental marketplace.

Senator BENNETT. There are some historians who suggest that the American oil embargo prior to World War II had a significant impact on the causing World War II. I am not well enough versed on the issues that faced Franklin Roosevelt in those days to try to second guess him on that decision, but I think there are some parallels here, and I do not think that we are faced with the militarist regime of the kind that the Japanese were prior to Pearl Harbor. But we seem to be about to do the same kind of thing as we did in the oil embargo against the Japanese.

Mr. BRODY. Well, I would add, we would have a lot more to benefit. I do not know the details of the oil embargo against the Japanese, but certainly the United States has a tremendous amount to benefit from continuing MFN status.

I do not know if that was the case with the Japanese, but this time not only will be punish China, we will certainly punish ourselves.

Senator BENNETT. Yes, I also do not know what we had to benefit, but I do know that the Japanese war planners came to the conclusion that without oil their economy would come to a complete halt and, therefore, they had to obtain oil by force of arms. And it was part of the justification given to Emperor Hirohito for their launching their military adventurism. I am wondering if there are any parallels here.

Mr. BRODY. I just add irony. An interesting irony is that as we talk about Shenyang today, Shenyang was the site in China that Japan put their sites on for the very reason of its energy reserves and its industrial capacity, and Shenyang was the capital of the Manchurian Dynasty. And it was the center from which Japan essentially managed Northeast China during the World War II period. So, there are interesting connections here.

And I would really hope that we would proceed cautiously and proceed in bringing our world community together. And I think that former President George Bush understood very well the relationship of the United States and China, and I commend him for strengthening our relationships with that country. And I feel that we should definitely proceed with caution in this regard.

Senator BENNETT. I thank you all. We are preaching to each other now, and trading platitudes back and forth, and that is always a good signal that the time has come to bring the hearing to an end.

Thank you all. The hearing is adjourned.

[Whereupon, at 12:06 p.m., the hearing was adjourned.]

APPENDIXES

APPENDIX I

Responses to Additional Questions

DEPARTMENT OF ENERGY,
CONGRESSIONAL AND INTERGOVERNMENTAL AFFAIRS,
Washington, DC, May 6, 1994.

Hon. J. BENNETT JOHNSTON,
Chairman, Committee on Energy and Natural Resources, U.S. Senate, Washington, DC.

DEAR MR. CHAIRMAN: On May 5, 1994, we sent you the answers to 39 questions from the March 16, 1994 hearing.

Enclosed are the answers to the 11 remaining questions submitted by you and Senators Wallop and Murkowski to complete the record.

If we can be of further assistance, please have your staff contact our Congressional Hearing Coordinator, Barbara Campbell, on (202) 586-8238.

Very truly yours,

WILLIAM J. TAYLOR, III,
Assistant Secretary.

[Enclosures.]

DEPARTMENT OF ENERGY,
CONGRESSIONAL AND INTERGOVERNMENTAL AFFAIRS,
Washington, DC, May 5, 1994.

Hon. J. BENNETT JOHNSTON,
Chairman, Committee on Energy and Natural Resources, U.S. Senate, Washington, DC.

DEAR MR. CHAIRMAN: On March 16, 1994, William H. White, Deputy Secretary of Energy, testified regarding the domestic and international implications of energy demand growth in China and the Pacific Rim.

Enclosed are the answers to 39 of the 50 questions submitted by you and Senators Wallop, Murkowski and Bumpers. The remaining 11 answers are in the clearing process and will be forwarded to you as expeditiously as possible.

If we can be of further assistance, Please have your staff contact our Congressional Hearing Coordinator, Barbara Campbell, on (202) 586-8238.

Very truly yours,

WILLIAM J. TAYLOR, III,
Assistant Secretary.

[Enclosures.]

RESPONSES TO QUESTIONS FROM SENATOR JOHNSTON

Question 1. What are the principal environmental impacts associated with increased energy demand in the Asia-Pacific Region?

Answer. Rapid growth in the Asia-Pacific Region has already led to a number of serious problems. A number of these problems are well-documented in the report by Senator Johnston on the January 1994 trip he and six other Senators took to this Region. Rapid growth is expected to continue, which will be reflected by increased energy demand. If care is not taken, existing problems could get worse and new problems will be created.

Countries in the Asia-Pacific region use a variety of energy sources. Coal and oil now account for the largest share of primary energy consumption. The overall energy balance varies by country. For example coal is the dominant source of energy in China and India; other countries have more diversified fuel supplies.

China stands out as the largest contributor to the region's air pollution with about two-thirds of the total SO_2 emissions and roughly half of the NO_x and CO2 emissions. Because of this, and expected rapid economic growth in China, any strategy to reduce emissions of these gases in the Asia-Pacific region will have to include increased attention to reducing emissions in China.

Between 1990 and 2010, primary consumption of all energy source is projected to increase rapidly throughout the Asia-Pacific region. Consumption of coal is expected to double; oil use will also nearly double and natural gas use will nearly triple, although this varies by country. Increased energy demand could have potentially large local and regional environmental impacts, but mitigating measures are available to these countries and they should be encouraged to use them.

Relative to global pollutants, projected increases in fossil fuel use translate to approximately a doubling of CO_2 emissions over the next 20 years in the Asia-Pacific region. China is expected to account for much of this, with about 30 percent of worldwide CO_2 increases during the next 20 years, over half of which is expected to be from coal use.

Coal. Coal is used widely for electric power generation and industrial applications. It is also used extensively in China and India for domestic needs, including cooking and heating. In China, energy-related environmental problems largely reflect the dominance of coal in the national energy balance. This is also true for India, although to a lesser extent.

Potential environmental impacts are associated with increased coal use, which in many cases will have minimal controls. Potential air pollution problems include emissions of particulate matter, sulfur dioxide and nitrogen oxides. Power plants are a major source of these pollutants, but these emissions can be significantly reduced through use of clean coal technologies for new power plants and retrofits to existing facilities. Ash must also be disposed of properly for it not to cause solid waste and water pollution problems. Coal mining will also increase and must follow good environmental and reclamation practices in order to avoid local impacts. Environmental regulations in OECD countries and some non-OECD countries in the Asia-Pacific region limit air emissions from power plants. Many of these countries have expressed an interest in using clean coal technologies for new power plants and industry. Measures for reducing pollution from coal use in the domestic sector include replacement of old appliances with more efficient ones, increased use of coal briquettes, and displacement of coal stoves and heaters by those using natural gas and electricity.

Oil. In OECD countries in the region (Japan and Australia), growth in oil demand is expected to come mainly from the transportation sector, while in non-OECD countries demand will increase in both transportation and industry sectors. As the number of vehicles increases, emissions of nitrogen oxide and volatile organic compounds (YOCs) could increase significantly, contributing to urban air pollution, especially in densely populated cities. Catalytic converters are the principal technology used to control vehicular emissions of nitrogen oxides. Increased use of oil in the industrial sector could also lead to increased VOC emissions. VOCs and nitrogen oxides are believed to be primary precursors of tropospheric ozone; a greenhouse gas produced when they react.

Natural Gas. Gas is used for a variety of purposes including electric power generation and in industry. While gas is generally considered to be a "cleaner" fuel compared with oil and coal, potential pollution associated with increased use of gas is similar to that for the other fossil fuels in general. The need for emissions controls is determined by each country's environmental regulations and may not be stringent enough to require controls where gas is involved.

Hydropower. Hydropower is expected to increase significantly throughout the region with most of the increase occurring in China and South Asian countries. Potential environmental problems associated with development of hydropower include siltation of rivers and lakes, erosion, loss of productive farmland and dislocation of local populations. In China, impacts from hydroelectric power are, after those associated with coal use, the most important energy-related environmental problem.

GREENHOUSE GAS EMISSIONS

Question 2. What level of greenhouse gas emissions are likely from China in the future?

Answer. Rapid economic growth and heavy dependence on coal could make China the largest contributor to the growth in world greenhouse gas emissions between now and 2010. For example, the Energy Information Administration in its *International Energy Outlook 1993* estimates that carbon emissions in China could grow from 646 million metric tons in 1990 to 1,187 million metric tons by 2010. Carbon emissions are expected to grow twice as fast in China (3.1 percent per year) as in the world as a whole (1.5 percent per year) between 1990 and 2010. Should substantial changes occur in country or international policies dealing with energy efficiency or carbon emissions, these estimates could change dramatically.

Question 3. How interested is China in environmental protection, given the importance they place on economic growth and increasing energy production?

Answer. In our experience, there are several signs of a Chinesecommitment to environmental protection, particularly when it contributes to economic growth. For example, they have shown considerable interest in improving their energy efficiency as confirmed by their interest in the Department's integrated gasification combined cycle initiative and the establishment of the Beijing Energy Efficiency Center. In addition, the Chinese are currently completing their plan for implementing Agenda 21, covering a range of environmental problems, which they signed at the Earth Summit held in Rio de Janeiro in 1992. The development of an implementation plan by the Chinese for the Agenda 21 agreement is a positive sign of their commitment to environmental protection.

China, like all countries, must balance a number of economic, energy and environmental goals. They are currently undergoing a rapid economic expansion, which has an associated requirement to expand energy supplies. If the Chinese are to continue to mitigate the environmental implications of rapid economic growth, they will need the technical expertise of developed countries and the active encouragement of countries like the U.S. The U.S. is assisting China in meeting its environmental needs through initiatives such as the U.S. Country Studies Program and the Assisting the Deployment of Energy Practices and Technology Program.

Question 4. China's current and projected levels of environmental pollution pose profound human health risks. Isn't this an important human rights issue?

Answer: The Department of Energy has been consistent in its support for increased use of cleaner-burning fuels both within the U.S. and globally, and we believe this will reduce health risks. However, the issue of defining what comprises a human rights issue is not within the jurisdiction of the Department of Energy.

Question 5. What types of actions should be undertaken to reduce China's greenhouse gas emissions in the future? Is it possible to change China's path?

Answer. China is a developing country rich with energy resources and therefore has a wide range of energy options. To our knowledge however, only two studies are considering the CO_2 emission reduction technology options for China. They are funded by the Asian Development Bank (ADB) and the World Bank. Neither final report is currently available.

The U.S. Country Studies Program sent a Federal interagency team to China from March 7-17, 1994, to assist them in the preparation of a proposal to the Program. The study, if funded, would analyze a range of technology and policy options for China. The draft proposal includes the analysis of several policy options that have not previously been officially assessed by China.

More importantly, the study will be officially adopted as a Chinese country study under the rules of the U.S. Country Studies Program. Most of the analytic effort will be undertaken by Chinese analysts. Acceptance by the Chinese Government of the results of this study should contribute to Chinese policy making with respect to global climate change. It should therefore have some effect on the Chinese energy path in the future. The study is expected to take approximately two years to complete after the award of the U.S. assistance.

Question 6. To what extent will DOE's clean coal technology initiatives help to address the adverse environmental impacts associated with increased coal use?

Answer. China is very strongly dependent on coal for energy, especially electric power generation. They currently use coal in a very inefficient and dirty manner. Their baseload power systems operate at an efficiency of about 25% and very few are equipped with modern emission control systems.

The proposed international initiative will include facilitating the deployment of an Integrated Gasification Combined Cycle (IGCC) Power Plant that will provide the Chinese the cleanest and most efficient technology available for producing power from coal. First generation IGCC systems have an efficiency in the 42% range. Advanced systems will have efficiencies over 50% (more than double the efficiency of current power plants: which would also reduce CO_2 emissions by 50%). IGCC systems reduce SO_2 emissions by over 98%, NO_x emissions by over 90%. Solid waste would be reduced proportional to efficiency gains and further reduced by recovering sellable by products.

The eventual total national benefit to China and the reduction in emissions of local and global interest will depend on the extent to which IGCC systems are deployed to meet their future energy needs.

82-926 O - 94 - 3

Question 7. Given that China is going to emphasize coal production for the foreseeable future, shouldn't the U.S. program focus on clean coal technology?

Answer. Yes. The proposed International Initiative for clean coal will focus on coal use for the regions with a large percentage of coal use—China and Eastern Europe.

NUCLEAR POWER

Question 8. To what extent can China rely on nuclear power as a means of reducing greenhouse gases?

Answer. China has a significant indigenous uranium and enrichment capability and is developing an indigenous nuclear powerplant industry. U.S. industry estimates up to 18 plants could be started in the time period of 1993 to 2007. This deployment rate would depend on availability of adequate capital resources.

ENERGY EFFICIENCY/RENEWABLES/METHANE

Question 9. What role will energy efficiency and renewable energy programs play in reducing environmental impacts of energy use in China?

Answer. Because of China's current heavy reliance on coal to produce electric power, the utilization of both energy efficiency and renewable energy technologies hold great promise to improve environmental conditions in the country. Energy efficiency will do this by reducing demand for electric power, and the utilization of renewables by producing power with clean, non-polluting electric power generation systems. China is the world's second largest emitter of greenhouse gases, producing considerably more carbon dioxide per unit of economic output than the U.S. Current emission levels are expected to triple or even quadruple by the year 2025, depending on whether or not China improves the efficiency with which it utilizes energy. Inefficient energy use in China not only heightens the risk of global climate change, but also harms the local environment and hampers economic development. Energy waste leads to power shortages that erode economic productivity, and to excessive coal combustion that causes severe air pollution and acid rain. DOEs energy efficiency efforts are centered in the Beijing Energy Efficiency Center, established with the assistance of the Department. The Center, staffed principally by Chinese citizens, serves as a focal point for disseminating information about U.S. energy efficiency technologies, and as a vehicle for analyzing the role that efficiency can play to improve energy utilization in China.

With respect to renewable energy technologies. there is a rapidly growing interest in their greater utilization in China. To a significant degree this is based on environmental concerns, with the result that many organizations at the national, provincial, and local government levels view renewables as a key ingredient in strategies to reduce pollution that emanates from power production and industrial facilities. In early 1994 a DOE-sponsored renewable energy mission to China took place, composed of representatives from U.S. industry and the National Renewable Energy Laboratory. While in China, the team met with a number of national government ministries, National People's Congress members, the Chinese Academy of Sciences, and provincial officials. Those meetings indicated a very strong interest in U.S. renewable energy technologies, both for the purpose of reducing emissions from conventional power plants, as well as to meet rising energy demands in remote areas of China where currently 150 to 200 million people have inadequate supplies of electrical energy. DOE and its renewable energy stakeholders are actively following up on the trade mission and, in collaboration with the U.S. Environmental Protection Agency, are exploring ways in which the work of the Beijing Energy Efficiency Center could be expanded to also deal with renewable energy matters.

Question 10. Given the obvious focus on growth, how effective are energy efficiency programs on the ground?

Answer. Energy efficiency programs are often the most cost-effective means of supplying energy services to buildings or industry—whether an economy is expanding or relatively stable. While China will need substantial investments in electricity generation capacity as its economy expands, investments in energy efficient heating, lighting, industrial process and other technologies could make existing capital go further by providing sore energy services for the same overall investment while producing environmental benefits.

CLIMATE/FRAMEWORK/JOINT IMPLEMENTATION

Question 11. What is the United States doing to help China under the framework convention?

Answer. At the request of the Government of China, the interagency United States Country Studies program, led by the Department of Energy, the Environmental Protection Agency and Agency for International Development, is considering assisting China in a study that would support their participation as a Party to the Climate Change Convention.

Recently, a Country Studies Project Team worked with representatives of several Chinese ministries, assisting them to develop their proposed study and associated budget. As a result of this team's visit to Beijing, an integrated analytical proposal has been completed by the Chinese study team and will be considered for funding later this spring by the Country Studies program. This Country Study will differ from previous studies by international organizations in that the analysis will be done almost entirely by Chinese scientists and analysts. Areas of study in the proposal include data collection, literature evaluation, methods development and adaptation to Chinese regions, workshops (including Ministerial workshops on Technical and Policy Mitigation Options), and simulations and modelling. The study is expected to include U.S. technical and financial assistance over a two year period. The results would be one basis of communications for the Chinese government to meet its obligations of the Climate Change Convention. The first Chinese communication will be due in March 1997.

Question 12. Is China considered a strong candidate for joint implementation projects? What is EPA doing to promote JI projects in China?

Answer. We consider China to be a very strong candidate for joint implementation projects, assuming that China chooses to participate in such international cooperation efforts. The Chinese delegation to the Intergovernmental Negotiating Committee (INC) on the Framework Convention on Climate Change has expressed concerns about the use of joint implementation under the provisions of the convention. The delegation has supported, however, the consensus of the Working Group of the INC that pilot projects such as the United States Initiative on Joint Implementation are useful efforts for the development of acceptable methods and criteria for joint implementation.

Recently, a Country Studies Project Team worked with representatives of several Chinese ministries, assisting them to develop their proposed study and associated budget. DOE, EPA and AID generally manage the Country Studies program. As a result of this team's visit to Beijing, an integrated analytical proposal has been completed by the Chinese study team and will be considered for funding later this spring by the Country Studies program. This Country Study will differ from previous studies by international organizations in that the analysis will be done almost entirely by Chinese scientists and analysts. Areas of study in the proposal include data collection, literature evaluation, methods development and adaptation to Chinese regions, workshops (including Ministerial workshops on Technical and Policy Mitigation Options), and simulations and modelling. The study is expected to include U.S. technical and financial assistance over a two year period. In addition, EPA and DOE have been reviewing a technology cooperation project between the State of Illinois and China.

Question 13. Several of the National Labs are actively involved with the Beijing Energy Efficiency Center. Does DOE see any other role for the Labs in technology transfer programs in the region?

Answer. DOE sees a potentially valuable role for the National Labs in international technology transfer in providing technology information, analytical assistance, and training. Establishing effective teaming of government officials, private industry and the laboratories will be the key to expanding these activities. For example, DOE established in FY 93 the Assisting Deployment of Energy Practices and Technologies Program (ADEPT) to assist developing countries with enhancing their policy-making and analysis infrastructure in support of sound energy and environmental policies. In FY 93, nine projects were funded, of which five were in Asia. They were:

Beijing Energy Efficiency Center—establishes an institution to focus on energy efficiency promoting policies, developing financing mechanisms, and opening a window on U.S. technology. This project is managed through Battelle, Pacific Northwest Laboratory (PNL) and Lawrence Berkeley Laboratory (LBL).

Biomass Ecosystem Data Development in Thailand—develops baseline data on ecosystems to be used in evaluating the environmental implications of biomass energy and agricultural options. This project is managed through Argonne National Laboratory.

Advanced Window Technologies in Malaysia and India—demonstrates the efficiency of advanced window systems in differing national circumstances. This project is managed through Lawrence Berkeley Laboratory.

Transportation Energy Efficiency and Environmental Performance in Indonesia, Thailand, and Chile—assesses the potential for improvement and development of alternative transportation energy policies. This project is managed through Oak Ridge National Laboratory.

Energy Efficient and CFC-Free Refrigerators in India—This project works with Indian refrigerator manufactures, U.S. refrigerator manufactures and the Indian TATA Research Institute to adapt U.S. technology to local conditions and to establish a market demonstration. This project is managed through Oak Ridge National Laboratory.

Except for the Thailand biomass project, each of these projects was also funded in FY 94. In FY 94, the ADEPT program received a number of new project proposals for Asia. Unfortunately, budget limitations prevented the Department from funding added projects. In FY 93, ADEPT had 51 million of reprogrammed funds; in FY 94, $750,000 was committed from the funding of the Office of Policy, Planning and Program Evaluation at the Department.

RESPONSES TO QUESTIONS FROM SENATOR WALLOP

GENERAL QUESTIONS

Question 1. What activities does the Administration propose to strengthen the services being provided by American missions and embassies in the Pacific region and elsewhere to actively support expanded trade in energy and environmental technologies?

Answer. The Department of Energy is supporting the Trade Promotion Coordinating Committee (TPCC) effort to develop and implement coordinated interagency programs to support U.S. firms in key markets, including those of the Pacific region. The following summarizes some of the key initiatives underway and those that are planned. The Department chairs an interagency group developing an electric power action plan for Indonesia. We led the interagency effort to develop an electric power initiative for China. We have taken private sector groups to Thailand, Taiwan, Indonesia, and Vladivostok, Russia to promote the use of U.S. energy technologies. We have requested funding for a demonstration project for China, which should help to disseminate U.S. clean coal technology throughout the region. As part of the interagency TPCC process, we are considering other smaller demonstration projects. We are planning five specialized trade missions related to electric power to China as well as technical conferences for Korea, Australia, and Indonesia. As in all programs that require resources to implement, DOE and all of the TPCC agencies must grapple with the question of how to strengthen the export program in the Pacific with shrinking budgets. Utilizing the personnel and financial resources of the 19 TPCC Agencies in a cooperative manner will certainly increase the productivity of the U.S. Government effort to support private sector competitiveness.

Question 2. Other countries such as France and Germany already are actively supporting expanded trade in energy and environmental technologies.

(a) What technical and financial support are our competitors obtaining from their national governments?

Answer. The governments of all major industrialized countries provide support similar to our own—support for research and development, export assistance programs and financing. People of good will can argue over whether foreign support programs are superior to ours without coming to any conclusions. However, the Department believes it is important to note that this Administration has undertaken a very active program of export promotion. Under the leadership of Commerce Secretary Brown, all of the Cabinet and Sub-Cabinet officials have agreed to form an advocacy network to use their contacts to support U.S. industry. The TPCC has begun the first-ever effort to develop creative, new ways of assisting U.S. industry. U.S. energy firms are very competitive. With the program the Administration has initiated, we believe substantial progress will be made in meeting the competition worldwide.

Question (b). What role is being served by their embassies? How does this compare with the services being provided by our embassy?

Answer. Their embassies provide the same kind of services that the U.S. State Department and Foreign Commercial Service provide. We can argue points of quality and quantity, and it is true that in some cases foreign governments provide substantially greater personnel resources to the commercial sections of their embassies. But again, we have some outstanding commercial officers in the U.S. embassies in the Asia Pacific region and we believe the *National Export Strategy* provides a firm basis for addressing these issues.

Question 3. Reportedly, clean coal technologies are not considered environmental technologies for the purpose of United States trade promotion. If so, what is the basis for this policy? What does the Clinton Administration propose to do to correct this situation?

Answer. The reports you have received are not accurate. Clean coal technologies are included in the TPCC Environmental Trade Working Group. For example, in carrying out the *Environmental Technologies Exports* report, DOE chairs a cluster group on Overseas Demonstrations which includes our clean coal technology demonstration plant proposals for China and Eastern Europe.

ENVIRONMENTAL TECHNOLOQY INITIATIVE

Question 4. What is the "Environmental Technology Initiative"? What is the statutory basis? What agencies are involved? What are its FY 94 and proposed FY 95 funding levels?

Answer. The "Environmental Technology Initiative" is a multi-year program developed and led by the Environmental Protection Agency (EPA), which will be conducted with other Federal agencies and the private sector. The Initiative, announced by President Clinton in his State of the Union speech on February 17, 1993, focuses on long-term research and pollution prevention to accelerate environmental protection while strengthening the marketplace for U.S. environmental technologies, both domestically and internationally. The FY 1994 scope of the Environmental Technology Initiative is described in the *U.S. EPA Environmental Technology Initiative: FY 1994 Program Plan* (EPA/542/K-93/OO3) dated January 1994. In FY 1994, a total of 73 specific projects were selected by the EPA's Innovative Technology Council for funding under the Environmental Technology Initiative.

In developing the Environmental Technology Initiative Program Plan for FY 1995, the EPA recently invited the Departments of Commerce, Defense, Energy, and other appropriate Executive agencies to name representative to participate in the working committee of the EPA Innovative Technology Council. These committees will be charged with recommending focus areas and evaluating FY 1995 project proposals.

The Environmental Technology Initiative builds on the existing statutory authority of the Environmental Protection Agency to conduct research and development, as illustrated in the following statutes:

- Clean Water Act—Section 104 (33 CFR 1254);
- Clean Air Act—Section 103 (42 CFR 7403);
- Comprehensive Environmental Resource, Compensation, and Liability Act (CERCLA)—Section 311 (42 CFR 9660);
- Resource Conservation and Recovery Act—Section 8001 (42 CFR 6981); and
- Safe Drinking Water Act—Section 300j-1 (42 CFR 300j-1)

Question 5. Why is the Initiative important to EPA or DOE activities in the Pacific Rim?

Answer. The Initiative focuses on key issues associated with successful deployment of U.S. environmental technologies. We have strong technical leadership in many environmental technologies. The Initiative provides the basis for a government/industry partnership that will benefit the rapidly growing countries of the Pacific Rim and the U.S. firms that are successful in helping these countries meet their environmental goals. The Initiative also responds to provisions contained within the Energy Policy Act of 1992 (section 1211, renewable energy, and section 1331, clean coal technology export) which support the export of environmental technologies.

Question 6. How will the Initiative help American companies to break into Pacific Rim and China energy and environmental technology markets?

Answer. The Environmental Technology Initiative is the first serious interagency effort to study the issue of commercializing environmental technologies and forming an industry/government partnership to successfully deploy those technologies abroad. It will give them access to foreign government decision-makers, standing beside U.S. government officials. It will improve their understanding of foreign business practices, provide a showcase for technically superior products, and access to expanded financial support.

CHINA RELATED QUESTIONS

Question 7. Given China's priority on economic development, what importance does China place on environmental protection?

Answer. In our experience, there are several signs of a Chinese commitment to environmental protection, particularly when it contributes to economic growth. For example, they have shown considerable interest in improving their energy efficiency as confirmed by their interest in the Department's integrated gasification combined cycle initiative and the establishment of the Beijing Energy Efficiency Center. In addition, the Chinese are currently completing their plan for implementing Agenda 21, covering a range of environmental problems, which they signed at the Earth Summit held in Rio de Janeiro in 1992. The development of an implementation plan by the Chinese for the Agenda 21 agreement is a positive sign of their commitment to environmental protection.

China, like all countries, must balance a number of economic, energy and environmental goals. They are currently undergoing a rapid economic expansion, which has an associated requirement to expand energy supplies. If the Chinese are to continue to mitigate the environmental implications of rapid economic growth, they will need the technical expertise of developed countries and the active encouragement of countries like the U.S. The U.S. is assisting China in meeting its environmental needs through initiatives such as the U.S. Country Studies Program and the Assisting the Deployment of Energy Practices and Technology Program.

Question 8. It is the goal of the Chinese government to substantially increase per capita GNP from $290 in 1987 to $1000 by 2000, to $2000 by the year 2025. A very large portion of China's rural population— by one estimate 200 million rural residents— by another, 37 percent of the rural population— does not yet enjoy electricity. Thus, it is generally recognized that commercial opportunities exist to sell U.S. energy and environmental goods and services in China.

(a) What are the Administration's views about the importance of China as a trade partner?

Answer. China has the fastest growing economy and one of the biggest emerging markets in the world. China is one of the world's largest markets for planned infrastructure projects with plans to spend as much as $200 billion on infrastructure-related equipment, technologies, and expertise during the next seven years. We are exporting $8-9 billion annually to China, supporting 160,000-180,000 U.S. jobs. Our view is that there are great opportunities for U.S. business exports into China despite some restrictions on trade and investment.

Question 8(b). What activities has DOE or EPA undertaken to expand energy and environmental trade with China?

Answer. Last June, the Department led a delegation of 24 U.S. electric power equipment and service firms to China. These companies met with close to 500 representatives of China's utility industry. They prepared a report that they presented to the Secretaries of Energy and Commerce that identified the enormous opportunities and the impediments facing U.S. firms seeking to take advantage of those opportunities. The mission recommended several key actions to the U.S. Government to address these impediments. The Department has taken a strong leadership role in attempting to address the issues identified. Progress has been made in expanding Export-Import Bank programs, removing some key export controls, and increasing efforts to disseminate information on U.S. technologies. The State Department and the NSC are reviewing other U.S. policies affecting trade in the context of overall U.S. policy objectives.

Recently the U.S.-China Joint Commission on Commerce and Trade (JCCT) agreed to create several industry specific working groups, one of these covers energy technologies. The Department of Energy will co-chair this working group with the Department of Commerce. We are developing a work program that we will present to China that will focus on issues of mutual concern and promote a mutually beneficial increase in the trade relationship.

Question 8(c). What barriers exist to the expansion of our trade with China?

Answer. Over the last decade, for reasons of Human Rights and Nuclear Non-proliferation, restrictions and sanctions that affect the energy industry have been imposed by the U.S. or remain under consideration. The most important of these provisions are:

The Agency for International Development is prohibited from operating in China;

The Trade and Development Agency and the Overseas Private Development Corporation are prohibited from obligating foreign assistance funds to China without a Presidential Waiver;

The Export-Import Bank is restricted by statute to a $50 million loan ceiling;

Recent policy restricting the use of tied-aid credits;

Until recently, a policy restriction on the export of balance-of-plant equipment for Chinese nuclear power-plants;

A condition for Presidential certification during the Congressional approval of the U.S.-China agreement for nuclear cooperation, which prohibits the sale of equipment for a nuclear power-plant until certification can be made;

Export controls on strategic commodities and technologies, such as high speed computers, numerical control machine tools, nuclear materials, and other dual-use commodities; and

The requirement for an annual determination of most favored nation tariff status, based on progress in human rights.

Question 9(a). In your opinion, is China prepared to upgrade the efficiency of existing coal fired power plants in response to local and regional environmental concerns?

Answer. Indications we have thus far suggest that China is prepared to upgrade the efficiency of existing coal power plants. For example, efficiency improvements are planned at many existing plants, especially in units of less than 20 MW. A joint venture between Westinghouse and Long Yuan Power Technology will undertake retrofits to improve steam turbine efficiency. Further incentive for making such improvements could result from plans by the Chinese Environmental Protection Agency to revise their Clean Air law and make it more comprehensive (i.e., cover SO_2 and NO_2 besides just particulate matter) and more enforceable.

Question 9(b). In your opinion, is China prepared to build power plants that are responsive to international environmental concerns such as climate change and acid rain?

Answer. We believe that China will build environmentally conscious powerplants, but several years will be needed to verify this. Factors cited in the previous answer support such a commitment. In addition, DOE has proposed in its FY 1995 budget to support construction of a high efficiency powerplant in China. The Chinese are very supportive of this project, and hope to use data from it to develop a strategy for future powerplant deployment.

A primary impediment to developing countries such as China to using state-of-the-art environmental technology is that such technology is frequently more costly. It is hoped that growing environmental awareness in China, coupled with cost reductions as new technologies (such as integrated coal gasification combined cycle) mature will result in the deployment of environmentally responsive power plants.

Question 10. What is China's policy regarding participation and investment by American business in their energy infrastructure?

Answer. The Chinese Government encourages foreign investment in the energy sector on the basis of mutual benefit and profit. Foreign investment in all energy sectors, including renewable energy sources and capture and use of coalbed methane, is now urgently needed for China and other Asian countries to cope with enormous power demand increases projected over the next decade.

The Chinese Government has recently moved toward three important policy positions with respect to foreign investment in power-plants. First, while the Chinese government understands that foreign developed power projects would be more costly than conventional Chinese power-plants and that rate tariffs would have to reflect that fact, electricity prices will not be allowed to rise to a level above the consumer's capability to pay. Second, the Chinese partner in an investment will take the majority share in any project because, in the Government's view, the power industry is fundamental to the infrastructure. Third, the Chinese do not believe a risk factor should be included as a cost in the electric power business. The Chinese believe that American companies often include inflated risk factors in their calculations of internal rate of return for project development. DOE, DOC and other U.S. Officials are discussing these issues with Chinese officials in an effort to persuade them to consider the impact these policies might have on American investors now negotiating contracts in China.

Question 11. When an American manufacturer competes on a project in China, are American manufacturers competing against Chinese manufacturers? Are the Chinese manufacturers being subsidized?

Answer. In some cases American equipment suppliers may compete against Chinese manufacturers, e.g., when open bidding is required by multilateral development banks. For most equipment purchases, open bidding is not practiced. The Chinese identify equipment for which they do not have the required technology and for which they are willing to allocate hard currency. In some cases, U.S. firms provide equipment from joint venture manufacturing operations in China.

In a non-market economy, the question of whether Chinese manufacturers are being subsidized is a very complex question, and no completely satisfactory methodology exists for determining if a subsidy exists or how large that subsidy might be. The Department of Commerce has developed a methodology and provides analysis on this issue.

Question 12. Witnesses have advocated that the activities of the Trade Development Program and the Overseas Private Investment Corporation be resumed in China. Others have advocated expansion of EXIM Bank in lending in the area.

(a) Would you support linking a portion of such technical and financial assistance to the purchase of American goods and services? If not, why not?

Answer. The historical record adequately demonstrates that when the full services of EXIM Bank, TDA, and OPIC are applied in a country, the success of American firms significantly improves and the value of our exports increases. As soon as we can resolve our non-trade issues with China, it would be beneficial to make available the full services of these agencies in support of trade with China. In the past, in order to offset measures of other governments, provisions have been made for requiring the purchase and American goods and services.

Question 12(b). What types of Federal government support would counteract the trade practices of other nations?

Answer. Under the framework of the Trade Promotion Coordinating Committee, a tied-aid fund of $100 million for FY 94 and $150 million in FY 95 will provide support that counteracts other nation's practices. The Department also believes that overseas demonstration projects, using Federal assistance, are extremely valuable in exhibiting the superiority of American technology. Demonstration projects such as clean coal technology, photovoltaic rural electrification, and environmental restoration technologies advocated by the Department are good examples.

Question 13. What are the practices of the Chinese Government regarding the protection of intellectual property?

Answer. China has recently improved its legal regime regarding the protection of intellectual property rights by implementing a full range of intellectual property legislation and regulations that address copyright, trademarks, patents, and unfair competition. However, China's failure to enforce its new legislation is of major concern to the United States. For instance, the 1994 National Trade Estimates Report (NTER) of the United States Trade Representative (USTR) states that, according to U.S. industries, China's failure to date to enforce its new copyright legislation has contributed to annual U.S. losses of more than $600 million.

The 1994 NTER includes a discussion of the trade activities of the People's Republic of China, including a discussion of China's intellectual property rights protection. According to the NTER, pursuant to the 1979 United States-China Trade Agreement, the United States and China agreed to offer copyright, patent and trademark protection equal to the protection correspondingly accorded in the other country. For several years, the two countries have discussed detailed methods for improving China's protection of intellectual property. These discussions covered copyright, patent, trade secrets, and unfair competition and trademark laws, and enforcement of these laws. After making little progress in these discussions, on April 26, 1991, pursuant to the "Special 301" provisions of the 1968 Trade Act, the USTR identified China as a "Priority Foreign Country." The USTR then initiated a "Special 301" investigation in May 1991 on China's intellectual property rights practices. That investigation was resolved on January 17, 1992, when the United States and China signed a Memorandum of Understanding (MOU) on Intellectual Property Rights. In this MOU, China has committed to providing improved protection for U.S. inventions and copyrighted works, including computer software and sound recordings, as well as trade secrets. China has made progress in the implementation of the MOU by implementing new legislation and regulations to afford intellectual property rights protection. China has also sought entry into the General Agreement on Tariffs and Trade and become a member of the Patent Cooperation Treaty. However, the enforcement of its new legal regime governing the protection of intellectual property rights is sporadic at best and virtually non-existent with regard to copyrighted works. The USTR is now in the process of negotiating with the Chinese to have China enforce the laws that have been implemented, especially with regard to foreigners.

This response was prepared with input from the United States Trade Representative and the United States Patent and Trademark Office. Should the Committee require additional information or have additional questions concerning China's practices regarding the protection of intellectual property, the Committee may contact Tom Robertson, an Intellectual Property Attorney with the Office of General Counsel, United States Trade Representative, at (202) 395-6800.

CLIMATE CHANGE (JOINT IMPLEMENTATION)

Question 14. What activities are EPA and DOE doing in China under the Framework Convention on Climate Change?

Answer. At the request of the Government of China, the interagency United States Country Studies program, led by the Department of Energy, the Environmental Protection Agency and Agency for International Development, is considering assisting China in a study that would support their participation as a Party to the Climate Change Convention.

Recently, a Country Studies Project Team worked with representatives of several Chinese ministries, assisting them to develop their proposed study and associated budget. As a result of this team's visit to Beijing, an integrated analytical proposal has been completed by the Chinese study team and will be considered for funding later this spring by the Country Studies program. This Country Study will differ from previous studies by international organizations in that the analysis will be done almost entirely by Chinese scientists and analysts. Areas of study in the proposal include data collection, literature evaluation, methods development and adaptation to Chinese regions, workshops (including Ministerial workshops on Technical and Policy Mitigation Options) and simulations and modelling. The study is expected to include U.S. technical and financial assistance over a two year period. The results would be one basis of communications for the Chinese government to meet its obligations of the Climate Change Convention. The first Chinese communication will be due in March 1997.

Question 15. (a) What is China's position on joint implementation under the United Nation's Framework Convention on Climate Change?

(b) Doesn't China endorse the position of the G-77 in opposition to joint implementation?

(c) What is the Administration doing to promote joint implementation projects in China? What arrangements are being made regarding any greenhouse gas reduction credits?

Answer. The Chinese delegation, at the Ninth Session of the Intergovernmental Negotiating Committee (INC), supported the position of the Group of 77 that initial joint implementation activities under the Framework Convention should involve only the countries listed in Annex I of the convention. Their major stated concern in doing this was to assure that obligations of the Annex I countries under the convention were not transferred to the developing countries.

The Chinese delegation joined the consensus of the meeting that national pilot programs such as the United States Initiative for Joint Implementation (USIJI) were appropriate in order to gain experience and to facilitate the later development of the criteria for joint implementation under the convention.

The Administration has not initiated its program but is now establishing the Evaluation Panel. The scope of activities by the Panel and its supporting organizations will be developed by this Panel. Some projects have been suggested by private developers that would be in China and may qualify as projects under the USIJI.

Because the USIJI has not begun operations there are no decisions made regarding assignment of greenhouse gas emission reduction credits as part of this program. At the Ninth Session of the INC, the delegations agreed that the present obligations of the convention were such that no assignment of credits was needed at this time in the national pilot projects.

CHINA RELATED QUESTIONS

Question 15. To what extent can, or does, China intend to rely on nuclear power as a means for meeting its electricity needs or reduce greenhouse gas emissions?

Answer. China has a significant indigenous uranium and enrichment capability and is developing an indigenous nuclear powerplant industry. U.S. industry estimates up to 18 plants could be started from 1993 to 2007. While China has stated plans to build nuclear powerplants as rapidly as possible, they will most certainly have to rely on coal and other energy sources in the less industrialized regions.

China recognizes the difficulty in realizing economic development while protecting the environment and has chosen to formulate their Agenda 21 to meet elements of the Rio United Nations Conference on Environment and Development.

CLIMATE CHANGE (JOINT IMPLEMENTATION)

Question 16. What is China's position regarding the development of a Climate Action Plan itself? Isn't it opposed to binding obligations on developing countries?

Answer. The Chinese delegation has strongly asserted that there should be no obligations placed on developing countries to reduce their net greenhouse gas emissions. China, by ratifying the Convention, accepted the obligation to provide a first communication to the Parties of the Convention by March 1997 on how it is meeting or is planning to meet its obligations under the convention. This communication is to include reporting on its policies and programs to limit net emissions. This communication is equivalent to what is being called by many countries their National Climate Change Action Plan.

GENERAL

Question 17. According to your testimony, the Trade Promotion Coordinating Committee provides the framework for aggressive support by this Administration.

What have been the activities of the TPCC under the Clinton Administration? When, and on what subjects, has the TPCC met?

Answer. The TPCC met on a very intensive schedule to develop the National Export Strategy, which was delivered to Congress last September. It has continued to meet on a regular basis to implement the programs identified in the NES. Most recently, the TPCC has been meeting to develop a trade promotion strategy for Indonesia, which was selected to provide a case study of strategies and tactics that provide the most effective approach to supporting U.S. firms in developing country markets. The TPCC also has been meeting on trade finance issues, developing an advocacy network, and improving the quality of commercial reporting.

Question 18. The 1992 Energy Policy Act envisions a more pro-active role for the Federal government in promoting exports of American manufactured energy technologies, particularly energy efficiency, solar and renewable, and clean coal technologies. DOE is given specific responsibilities working through the Trade Promotion Coordinating Committee to promote the export of domestic energy resource technologies to developing countries.

(a) Do you support the use of the Trade Promotion Coordinating Committee to promote the export of domestic energy resource technologies to developing countries, generally?

Answer. Yes, it is the only way to effectively utilize all of the resources available to the Administration to support exports.

Question 18(b). What have been the energy and environmental activities of the TPCC under the Clinton Administration? When, and on what energy or environmental subjects, has the TPCC met?

Answer. The TPCC selected energy as one of the key sectors to be supported by interagency resources. One of the first efforts has been the development of an export action plan for increasing the U.S. share of Indonesia's electric power equipment and service market. We expect that this plan, which is being developed by an interagency TPCC subcommittee under DOE leadership, will be presented to Secretary Brown and the TPCC membership by the end of April 15.

The TPCC Environmental Trade Working Group has established 7 cluster groups to carry out the recommendations of the *Environmental Technologies Exports: Strategic Framework for U.S. Leadership* report issued November 22, 1993. The Working Group meets monthly to review progress on implementing the recommendations.

Question 18(c). What else can the Federal government do to be more pro-active in the promotion of U.S. technologies?

Answer. The Administration, the TPCC and the Department have begun implementing a strong export promotion program. Within that program, energy is a priority sector. We need to be—and intend to be—more effective at implementing programs with long-term benefits. You can not go into a country the day before a bid award, lobby for a U.S. company and expect to be successful. We need to demonstrate a long-term commitment to working with foreign energy officials on the basis of mutual respect and mutual benefit. We need to do a better job of making certain that the benefits of U.S. technologies are well understood and that foreign officials are aware of U.S. technology options. We are requesting money for a demonstration project in China. As a result of a demonstration project in Poland, which yielded a commercial sale of the technology for another facility, we know that these projects promote U.S. exports and that the benefits extend beyond the demonstration project itself. Understanding the necessary constraints on budget expenditures, I would like to see more carefully selected demonstrations of key energy technologies.

Question 19. The 1992 Energy Policy Act sets forth established DOE programs to promote the export of already commercialized clean coal technologies, as well as renewable energy and environmental technologies. The Congress intended that these provisions to govern the related trade promotion activities of the Department of Energy.

What are DOE's plans for implementation of these programs?

Answer. The plans for implementing sec. 1332 of the Energy Policy Act involves three parts:

1. Provide technical assistance to foreign governments and U.S. export agencies, as well as U.S. firms. This is accomplished through many means: trade missions, reverse trade missions, workshops, third party review of feasibility studies, providing technical experts to review projects for U.S. credit agencies and multi-lateral banks, participating with other government agencies on the development of export strategies.

2. Providing minimal assistance (10-25 percent) to reduce the risk for two or three "showcase" projects to convince the largest potential markets (China and Eastern Europe) of the benefits of U.S. clean coal technology. The President's FY 1995 Budget includes $20 million for this.

3. Provide a pool of funds to reduce industries' front-end risk incurred for project definition activities, especially for moderately capitalized companies. The project definition costs are five times those for a domestic project. These include the engineering sufficient to develop a cost estimate in the range of plus 15 percent to minus 5 percent accuracy, development of supply and sales agreements, and identification of risks and the necessary mitigation of these risks.

This three step approach emerged from numerous consultations with U.S. industry and trade groups and are mechanisms being used by other governments to support their industries. The Administration is seeking a reprogramming of Clean Coal funds in FY 95 and FY 96 to implement the international program.

Question 19(b). Are DOE and EPA coordinating their respective export programs?

Answer. Yes, we are working closely with EPA on the Environmental Technology Initiative and in other areas. One example of this cooperation is the program under way in the EPA/DOE co-chaired Gazprom Working Group. Our two agencies are working with the Russian Gazprom organization to address issues associated with the safety and efficiency of gas pipelines. The Working Group, in cooperation with U.S. firms, has identified 20 projects that have the combined benefit of demonstrating U.S. technologies and addressing Russia's pipeline leakage problem.

Question 20(a). Do you support the use of the TPCC's Coal Subgroup for the promotion of Clean Coal Technology exports to developing countries such as China?

Answer. Yes, we do support use of a group such as the Coal Subgroup, which has proven effective since its inception. There have been recent organizational changes in the TPCC, and actions are underway to make this group even more effective.

Question 20(b). With respect to clean coal technologies, the Energy Policy Act establishes a Subgroup of the Department of Commerce Trade Promotion Coordinating Committee to be chaired by the Secretary of Energy. The EP Act also authorizes the Secretary of Energy, through AID, to provide financial assistance to American business when they compete for energy technology projects in foreign countries.

What are the Administration's plans for helping American companies compete for foreign clean coal technology projects?

Answer. The Clean Coal International initiative is designed to assist U.S. companies to compete through educating the foreign customers concerning performance and economics of the clean coal technologies; to support showcasing projects in China and Eastern Europe to convince the host countries of the benefits of the U.S. technology and to provide useful technical and economic information; and creating a pool of funds for cost sharing project definition studies to assist U.S. industry in developing project opportunities and to define these opportunities sufficiently to comply with the requirements of a lender.

Question 20(c). With respect to clean coal technologies, the Energy Policy Act establishes a Subgroup of the Department of Commerce' Trade Promotion Coordinating Committee to be chaired by the Secretary of Energy. The EP Act also authorizes the Secretary of Energy, through AID, to provide financial assistance to American business when they compete for energy technology projects in foreign countries.

What else can the Federal government do to be more pro-active in the promotion of Clean Coal technologies?

Answer. In addition to the proposed international clean coal initiative the Government plans to do some of the following to be more pro-active:

An organized effort throughout Government export credit agencies to promote clean coal exports.

Use influence to assist in obtaining sovereign guarantees.

Assist in the development of environmental legislation in the foreign countries.

Assist developing countries in creating accepted capital markets to expand the source of capital funds to the private sector.

Question 21. The Administration proposes to redirect a total of $100 million from the Clean Coal Technology Demonstration project account in Fiscal Years 1995 and 1996 to support the export of clean coal technologies. How does DOE propose to use this particular $100 million?

Answer. The President's FY 1995 Budget includes $20 million to start the international initiative. The plan for redirecting these funds in order to implement sec. 1332 of the Energy Policy Act involves three parts:

1. Provide technical assistance to foreign governments and U.S. export credit agencies, as well as U.S. firms. This is accomplished through many means: trade missions, reverse trade missions, workshops, third party review of feasibility studies, providing technical experts to review projects for U.S. credit agencies and multi-lateral banks, participating with other government agencies on the development of export strategies.

2. Providing minimal assistance (10-25 percent) to reduce the risk for two or three "showcase" projects to convince the largest potential markets (China and Eastern Europe) of the benefits of U.S. clean coal technology.

3. Provide a pool of funds to reduce industries' front-end risk incurred for project definition activities, especially for moderately capitalized companies. The project definition costs are five times those for a domestic project. These include the engineering sufficient to develop a cost estimate in the range of plus 15 percent to minus 5 percent accuracy, development of supply and sales agreements, and identification of risks and the necessary mitigation of these risks.

This three step approach emerged from numerous consultations with U.S. industry and trade groups and are mechanisms being used by other governments to support their industries. The Administration is seeking a reallocation of Clean Coal funds in FY 95 and FY 96 to implement the international program.

Question 22(a). According to your testimony, the Department has requested FY 95 funds to support a "show-case" demonstration project in China employing integrated coal gasification combined cycle technology.

What is the status of the identification or selection of this technology, the associated coal supply and the site for the generating plant? Will other clean coal technologies, such as fluidized bed combustion, be considered?

Answer. The Chinese have formed a team consisting of all the relevant government organizations that would be involved in the implementation of the IGCC project. The team is developing a pre-feasibility study to select the site, coal supply, capacity, candidate technologies and costs. The Chinese have identified IGCC as the most promising option for using coal in the future. Fluidized bed combustion is already in widespread operation in China.

CLEAN COAL TECHNOLOGY

Question 22(b). According to your testimony, the Department has requested FY 95 funds to support a "show-case" demonstration project in China employing integrated coal gasification combined cycle technology.

What is the nature of the demonstration project? For example, will the project be an integral part of an electric power system?

Answer. Yes. The project would involve financial assistance to a U.S. firm that participates in the design or construction of an Integrated Gasiflcation Combined Cycle (IGCC) plant ordered by North China Power Company. The IGCC plant would be an integral part of the North China Power Co., a government owned company.

Question 23. According to your testimony, DOE is currently co-funding the solar electric light fund in its efforts toward rural electrification in Western China.

a. What are the sources and amounts of FY 1994 and FY 1995 funding for this activity Center?

b. What organization is managing this account for the Department? What is their source of funding and its amount?

Answer. Funding for this activity center comes from the Photovoltaic Program within the Office of Solar Energy Conversion. FY 1994 funding was $30,000. Funding in FY 1995 is expected to be in the same range. The Photovoltaic Program is managing this activity/effort for the Department. The FY 1994 Photovoltaic Program budget is $73,958,000.

Question 24. According to your testimony, the Beijing Energy Efficiency Center was officially opened in December 1993. The Center intends to develop energy—efficiency policies and joint-ventures essential for China to make a transition to a more sustainable energy development pattern.

(a) What are the sources and amounts of FY 94 and FY 95 funding for the Center?

Answer. The following details the sources and amounts of FY 94 and FY 95 funding for the Beijing Energy Efficiency Center:

FY 95 Funding

U.S. DOE, Office of Policy, ADEPT Program	$200,000
U.S. EPA, Office of International Affairs	50,000
World Wildlife Fund International	30,000

FY 95 Funding (Anticipated but not committed)

U.S. DOE, Office of Policy, ADEPT Program	$200,000

Question 24(b). What organization is managing this account for the Department? What is the source of funding and its amount?

Answer. Battelle, Pacific Northwest Laboratory (PNL) and Lawrence Berkeley Laboratory (LBL) are managing the establishment of and support for the Center.

In FY 93, PNL received $125,000 and LBL received $75,000 of the $200,000 provided by DOE. Each lab established a subcontract with the Center for $70,000, with the balance retained by each lab for management. In FY 94, each lab received $100,000 from the Department and each will again provide $70,000 to the Center. The balance will again be reserved for management and expert assistance.

Question 24(c). Would you supply for the record a description of the activities that are going to be carried out by the Center, and those that will specifically help American business?

Answer. The Center promotes energy efficiency through four main activities: policy recommendations, private business development, training, and public education. All these activities contribute directly or indirectly to expanding opportunities for U.S. firms by helping create a market for energy efficient goods and services. The Center has already begun to provide direct assistance to U.S. firms in its first three months of existence by meeting with several company representatives to discuss potential joint-ventures in China. The Center intends to further assist U.S. firms by providing up-to-date information on China's energy efficiency technology priorities, targeted market studies, introductions to potential joint-venture partners, tours of potential sites, and information on how to do business in China.

Question 25. According to your testimony, a similar Center (Renewable Energy Center) is being established in China focused on renewable energy technologies.

Would you provide information comparable to question 24?

a. What are the sources and amounts of FY 1994 and FY 1995 funding for the Center?

b. What organization is managing this account for the Department? What is their source of funding and its amount?

c. Would you supply for the record a description of the activities that are going to be carried out by the center, and those that will specifically help American business?

Answer. A joint Department of Energy/Environmental Protection Agency proposal has been submitted to the Environmental Technology Initiative within the EPA. The requested amount in FY 1994 is $125.000. FY 1995 funding would be in the same range, with the goal of the Center becoming self-sustaining through user fees after several years.

If the proposal is funded, the Acid Rain Division in EPA, and the Office of Solar Energy Conversion in DOE will have responsibility for managing the effort.

The Renewable Energy Center would be a first point of contact for U.S. industry interested in the renewable energy market in China. The Center would keep all the latest market data accessible for U.S. business, as well as house up to date information on credible Chinese companies with which to partner where appropriate. In addition, assistance in navigating the Chinese economic system would be made available. This assistance would greatly reduce the lead time for U.S. businesses entering the Chinese marketplace, as well as enhance their chances of success.

Question 26. Will a similar center be established in China for clean coal technology? If not, why not?

Answer. The idea is being discussed and will be investigated in greater detail in the next six months.

Question 27. What similar centers are envisioned in countries other than China?

Answer. In FY 1995, two additional Sustainable Development Centers are planned which will be located in Costa Rica and Hawaii. These sustainable Development Centers, which will differ from existing energy efficiency centers located in Eastern Europe and Russia, will address both the energy efficiency needs and the use of U.S. renewable technologies within their geographical regions.

New centers

The center in Costa Rica will support the America's 21st Century Program in Latin America and the Caribbean. This program implements the regional strategy developed by the Committee on Renewable Energy Commerce and Trade (CORECT) to promote the export of U.S. renewable products and services. The center will assist U.S. firms seeking to do business in the region by providing local market information and by facilitating contacts with local decisionmakers.

The center in Hawaii will be developed in conjunction with the East-West Center and will provide a "gateway" to the Pacific market for U.S. firms. The center will provide information on the Pacific markets, business customs and institutional barriers to doing business in the region.

Existing centers

The Department currently supports Energy Efficiency Centers in Bulgaria, Poland, the Czech Republic, and Kiev (opening in late 1994). These centers are staffed by local personnel, who disseminate information on the capabilities of U.S. efficiency technologies to local decisionmakers. The centers have been recognized locally as areas of expertise and information. Although the centers were established with the help of the U.S. Government, they are expected to become self-sufficient within a reasonable period of time. The trend in the centers is positive; one-third to one-half of their funding now comes from sources other than the U.S. Government.

It is anticipated that the Energy Efficiency Centers in Eastern Europe will eventually transform into Sustainable Development Centers as the economies in the region begin to grow and environmental and renewable technologies become of greater interest.

In addition, the Department's Office of Fossil Energy and Office of Policy, Planning and Program Evaluation have received a grant from the Agency for International Development to establish an Oil and Gas Technology Center in Tyumen, Siberia. The primary objective of this Center is to provide a reliable source of information on U.S. oil and gas technologies, and to provide an interface between U.S. companies and potential Russian customers.

This effort was in response to the Gore-Chernomyrdin Commission's resolution to assist in the transformation of the Russian oil and gas sector industry to a market oriented, competitive industry.

RESPONSES TO QUESTIONS FROM SENATOR MURKOWSKI

Question 1. I am told that the Guandong province is commencing construction of three coal-fired projects this year to meet their rapidly expanding energy needs. Are any American companies involved in these projects, either to supply equipment or technology?

Answer. The U.S. Department of Commerce/TA Office of Major Projects confirms that three coal-fired projects are planned to commence construction this year in China's Guandong province. These projects are Ezhou, Yangzhou and Zhuhai. Advanced planning is in progress for an additional project at Ningbo. U.S. firms Foster Wheeler and Ebasco have been selected to provide boilers and engineering services for the Ezhou project. A number of U.S. firms are actively pursuing involvement in the other projects. These firms include General Electric, Foster Wheeler, Bechtel, ABB Combustion Engineering, Westinghouse, Black & Veatch and Mission Energy.

Question 2. I understand that the Guandong province is currently using coal from the northern provinces. Is Guandong considering importing coal in the future to meet its rapidly expanding energy needs? If so, what are its projected import demands?

Answer. DOE is not aware of any plans of the Guandong province to import coal. Instead the plans of the Chinese are to construct large mine mouth projects and transfer power by wire to the industrial areas.

Question. Has the APEC Working Group on Regional Energy Cooperation produced a policy paper? If so, what were the key findings? Please make a copy of the report available to my office when it is available.

Answer. Yes. Although the APEC Energy Working Group (EWG) has focused its initial efforts on developing energy programs, it has produced the attached paper entitled "Overview Policy Issues Paper". This paper was approved by the EWG at its October 1993 meeting and endorsed by the APEC Senior Officials (SOM) and the APEC Ministers in Seattle in November 1993.

The paper names three main policy objectives that the EWG could address:

- improving the contribution from the energy sector to economic growth;
- enhanced energy security, and;
- minimizing the adverse consequences of energy production and use.

The paper identifies the following sectors as warranting priority attention:

- electricity generation;
- transportation;
- rational energy consumption, and;
- natural gas.

A policy paper on rational energy consumption currently is being developed by the EWG.

Question. Has the private sector been given a formal role in the APEC Working Group on Regional Energy Cooperation? If not, are there any plans to include the private sector in future meetings of this APEC group?

Answer. Yes. Representatives of the Minerals and Energy Forum (MEF) of the Pacific Economic Cooperation Conference (PECC) have official observer status at APEC energy meetings, and participate actively. PECC/MEF is a tri-partite group that includes representatives from the private sector, government, and academia. In addition, senior government officials who participate in APEC regularly report to the MEF on APEC energy activities (see attached report made to the Sixth MEF, March 18, 1994).*

It is official policy of APEC in general, and the APEC Energy Working Group in particular, to promote private sector participation in its activities. Thus far, business/private sector representatives from the APEC member economies have participated in a variety of APEC energy events, including the May 1993 APEC seminar on demand-side management (DSM), the September 1993 APEC workshop on clean coal technology, and the March 1994 meeting of the APEC utility liaison group on DSM.

Increased private sector participation in APEC energy activities is anticipated. As an example, a listing of attendees at the May 1993 DSM seminar is attached.

ENVIRONMENTAL PROTECTION AGENCY,
LEGISLATIVE ANALYSIS DIVISION,
Washington, DC, August 23, 1994.

Hon. J. BENNETT JOHNSTON,
Chairman, Committee on Energy and Natural Resources, U.S. Senate Washington, DC.

DEAR MR. CHAIRMAN: Enclosed for insertion into the hearing record are EPA's responses to follow-up questions from the March 16, 1994 hearing on energy demand growth in China before the Committee on Energy and Natural Resources. I hope this information will be useful to you and members of the Committee.

If you or your staff have any questions regarding this information, please contact David Gardiner, Assistant Administrator for Policy, Planning and Evaluation at (202) 260-4332.

Sincerely,

THOMAS C. ROBERTS,
Director.

[Enclosure.]

RESPONSES TO QUESTIONS FROM SENATOR JOHNSTON

Question 1. What are the principal environmental impacts associated with increased energy demand in the Asia-Pacific region?

Answer. Increasing energy demand in the Asia-Pacific region is likely to he fueled through additional coal consumption, particularly in China. Coal consumption in China, for example, is forecast to grow from a current annual level of 1 billion metric tons to 1.5 billion metric tons in 2000 and could exceed 2 billion metric tons in 2020.

Growth in coal combustion will have serious adverse impacts for local air quality, particularly if pollution control devices are not employed. Coal combustion emits sulfur dioxide, nitrous oxides and particulates. If current trends continue and additional amounts of coal are burned in low-efficiency residential and industrial stoves and boilers, the local air quality in many regions of China could deteriorate substantially. In addition, increased coal combustion also has serious global environmental implications, due to the carbon dioxide emissions associated with combustion and the methane emissions associated with coal mining.

Question 2. What level of greenhouse gas emissions are likely from China in the future?

Answer. China's greenhouse gas emissions are forecast to increase dramatically in the future, primarily due to increasing coal combustion. The Asia Development Bank, for example, forecasts that carbon dioxide emissions associated with fossil fuel use in China could increase from 626 million metric tons of carbon in 1990 to 1354 million metric tons in 2020, under a "Business as Usual" scenario. This growth is projected because China is planing to fuel its explosive economic growth primarily through increasing coal production. Coal production liberates methane, a potent greenhouse gas, and coal combustion emits large amounts of carbon dioxide. By 2020, if not sooner, China will be the world's largest emitter of greenhouse gases. In addition, it will not be possible to stabilize greenhouse gas concentrations without the participation of China in programs to reduce emissions.

Question 3. How interested is China in environmental protection, given the importance they place on economic growth and increasing energy production?

---

*Retained in committee files.

Answer. China is committed to protecting the environment, but not at the expense of economic growth. Thus, the most effective strategies for addressing China's environmental impacts will be to focus on projects that make economic sense and are environmentally desirable. Coalbed methane and other alternative fuels, renewable energy technologies and energy efficiency are all examples of these types of projects. There are major business opportunities for U.S. companies in all of these areas.

Question 4. China's current and projected levels of environmental pollution pose profound human health risks. Isn't this an important human rights issue?

Answer. Clearly, the human health dimensions of China's environmental situation have a enormous impact on the quality of life of the Chinese population and thus can be viewed as a important factor in the overall human rights situation. China needs to take aggressive action to improve local environmental quality and contribute to the protection of our global environment. Cooperation with China on environmental protection can advance these objectives.

Question 5. What types of actions should be undertaken to reduce China's greenhouse gas emissions in the future? Is it possible to change China's path?

Answer. There are a variety of actions that can be undertaken to help China reduce its future greenhouse gas emissions and change its emission path. Possible actions fall into two general groups: actions that can improve the efficiency of energy use in China, enabling it to do more with less energy; and actions to increase supplies of clean-burning fuels, such as natural gas and renewables. Examples of specific types of actions that could reduce China's future greenhouse gas emissions include:

Programs to increase energy efficiency in all sectors, including residential, commercial and industrial;

Programs to increase the use of natural gas, which currently accounts for only 2 percent of China's energy supply;

Programs to increase the use of renewable energy sources, including wind, solar and hydropower; and,

Programs to increase the efficiency of coal combustion through the introduction of clean coal technologies and other efficiency improvements in the industrial, residential and commercial sectors.

Question 6. What role will energy efficiency and renewable energy programs play in reducing the environmental impacts of energy use in China?

Answer. Fully developing China's renewable energy resources and its opportunities to improve residential, commercial and industrial energy efficiency is of critical importance if China is to enjoy rapid economic growth in a environmentally beneficial manner. Ensuring reliable energy supplies will be essential to continuing China's rapid economic growth into the next century. Within this context, diversified energy supply, including natural gas and renewable energy sources in addition to coal, and efficient use of available energy provide a essential foundation for future economic growth.

Question 7. What is the U.S. doing to help China under the Global Climate Change Framework Convention?

Answer. The U.S. has several programs that are intended to help China under the Global Climate Change Framework Convention.

EPA's Office of Air and Radiation has a number of directed cooperative programs underway in China that will help China assess specific greenhouse gas emission reduction opportunities and identify projects that can reduce GHG emissions. These efforts include:

(1) A directed evaluation of opportunities to reduce methane emissions from various sources, undertaken in order to prepare a series of Reports to Congress requested under Section 603 of the Clean Air Act Amendments of 1990;

(2) Cooperation on development of energy efficient, ozone-friendly refrigerators which will reduce emissions of carbon dioxide in addition to protecting the stratospheric ozone layer;

(3) Informal support of ongoing efforts by the DP, World Bank and others to promote projects to reduce emissions and support various analytical efforts related to reducing greenhouse gas emissions; and

(4) Cooperation with the Department of Energy on efforts to establish the market potential and environmental benefits of various renewable energy technologies.

EPA's Office of Policy, Planing and Evaluation also has several ongoing activities with China, including:

(1) A three year study of methane emissions from rice paddies and potential management options for reducing the emissions;

(2) A study of the impacts of climate change on agriculture in China (recently published as part of a global study of agricultural impacts of climate change);

(3) A study of the effects of hot weather on mortality in two cities in China;

(4) Analyses of the impacts of sea level rise on China's coastal areas;

(5) An assessment of deforestation in China and the cost of mitigation options; and

(6) Model development and assessment of greenhouse gas emissions from the energy sector in China and options to reduce emissions.

Finally, China submitted a proposal to participate in U.S. Support for Country Studies to Address Climate Change in early 1994. U.S. Support for Country Studies, a two-year $25 million activity, is designed to provide financial and technical assistance to developing and transition countries in support of the Framework Convention on Climate Change. The $1.7 million China proposal requested U.S.technical support in four areas: (1) development of comprehensive GHG emission inventory; (2) evaluation of vulnerabilities to climate change, (3) assessment of potential mitigation and adaptation options to cope with climate change, and (4) assistance with development of national plans and reports in response to the Framework Convention process. A U.S. (EPA, DOE and DOC) delegation visited Beijing in March 1994, to work with the State Science and Technology Committee to refine the multi-ministry proposal for technical assistance. The U.S. delegation, alter consultation with the U.S. Embassy Science Office, recommended the proposal be supported. Funding decisions regarding support to China will be made following an interagency technical and programmatic review in FY 94.

Question 8. Is China considered a strong candidate for Joint Implementation Projects? What is EPA doing to promote JI projects in China?

Answer. The Administration views all Parties to the convention as eligible for joint implementation activities. China is the world's second largest emitter of greenhouse gases and those emissions are expected to grow rapidly in the next several decades due, in large part, to increased coal consumption. We believe that China's participation will be important to reaping the potential environmental and economic benefits of joint implementation in the long term.

EPA, with DOE, DOS and other federal agencies, is currently implementing the U.S. Initiative on Joint Implementation (USIJI), a pilot program under the President's Climate Change Action Plan. As the program is implemented, we will point towards facilitating proposals for JI projects in China and other countries with significant opportunities for large reductions in greenhouse gas emissions.

Question 9. What programs does EPA have underway in China and how do they relate to the Department of Energy's activities?

Answer. EPA has over 20 ongoing projects with China conducted through the US-PRC Environmental Protection Protocol, signed in 1980 under the umbrella US-PRC Science and Technology Agreement. Project activities fall under four annexes: Environmental Health, Pollution Control, Environmental Processes and Effects, and Environmental Management. In addition, we are undertaking exchanges and studies outside the Protocol. EPA's China program reflects the most productive aspects of EPA's international cooperation—cooperative research that advances both U.S. and PRC domestic environmental protection agendas, and cooperation that addresses the most critical global environmental concerns.

In the area of pollution prevention through expanded use of energy efficiency and clean energy sources, EPA has a few targeted programs underway in China. These programs have developed as components of activities undertaken as a result of the Clean Air Act Amendments and the Montreal Protocol. The principal activities include:

A project to improve the energy efficiency and reduce the ozone depletion potential of Chinese refrigerators. Basically, as China redesigns its refrigerators to reduce their ozone depletion potential in response to the Montreal Protocol, they can also take advantage of opportunities to make significant efficiency improvements;

A project to identify opportunities to reduce methane emissions from coal mines through the production of coalbed methane, a unconventional gas source pioneered in the United States. EPA is working closely with DOE's Office of Fossil Energy, building on a strong collaborative relationship developed between the agencies related to implementation of U.S. actions to promote coalbed methane mandated under the President's Climate Change Action Plan;

We are also working with DOE's Office of Renewable Energy and Energy Efficiency and others to assist China in expanding the use of renewable energy technologies and other energy efficiency projects, including support for the Beijing Energy Efficiency Center. We are working with the National Institute of Standards and Technology and external research programs to obtain research data on the properties of atmospherically safe refrigerants and refrigeration cycles.

In addition, we are working with the Department of Energy and other agencies on the Climate Change Country Studies Program. The Country Studies Management Team is working with China to initiate a comprehensive greenhouse gas emissions inventory and mitigation strategy.

All of EPA's pollution prevention programs which deal with energy efficiency or clean fuels are being coordinated with the appropriate offices at the Department of Energy. We have found such coordination improves the programs by providing additional perspectives and expanding access to, industrial groups in both the United States and China.

Question 10. How can EPA programs help U.S. companies develop business opportunities?

Answer. While EPA does not engage in the type of export assistance activities undertaken by the Department of Commerce, our international technical assistance, training, and other capacity-building programs enhance the demand for environmental technologies and services, thereby creating potential commercial opportunities for U.S. business and industry. EPA also works with the Department of Commerce and other federal agencies in making information on international environmental markets and needs available to interested parties.

Question 11. What are the main impediments to U.S. companies doing business in China?

Answer. From our experience, we have observed a fundamental impediment related to developing environmentally sound business opportunities in China: a lack of information related to the opportunity in China and in the United States. In may cases, the Chinese are unaware that there are attractive technologies available that could be profitably deployed and would reduce environmental impacts. Similarly, U.S. companies—particularly the smaller companies that dominate some of the niche energy and environmental markets—lack information on China's market conditions and do not have the financial resources necessary to develop the information and business connections independently.

While our programs cannot fully address these issues, we strive to facilitate the opening of information channels. For example, we provide information about various project opportunities to U.S. trade agencies and thus to U.S. companies. In addition, we have been quite successful in leveraging funding from the World Bank, the Global Environment Facility, and other organizations to support projects identified as a result of our involvement. U.S. companies have dominated the procurement for these projects.

Question 12. Isn't EPA stepping beyond its natural role in assisting China? Shouldn't the types of projects being undertaken by EPA more appropriately be implemented by DQE?

Answer. EPA's primary role is to assist its sister agency, the National Environmental Protection Agency, in identifying and implementing programs to protect the global and local environment. All of our activities to date have been grounded in obligations to collect information for Reports to Congress, in support of the Montreal Protocol, or due to requests from other agencies or Chinese counterparts. With respect to the EPA activities that affect China's energy supply and demand growth, China is extremely interested in pollution prevention and in the development of innovative programs that can reduce pollution through more cost-effective operations. In the U.S. we have found that both EPA and DOE can make important contributions to the success of programs that can profitably prevent pollution from energy use, because the agencies can reach different industry groups and undertake a variety of outreach, research and development and analytical activities. Many of these types of programs may also be of interest to China as it develops strategies to meet the twin goals of economic growth and environmental protection. In China, as in the U.S., both EPA and DOE can provide important assistance to China in improving its pollution prevention in the energy area.

Question 13. Why is EPA concerning itself with U.S. exports and international markets? Isn't that the responsibility of the Department of Commerce?

Answer. The transfer of environmental technologies and services plays a extremely important role in solving international environmental problems of concern to the United States. Recognizing that many of the relevant technologies are only available through commercial channels, EPA has a active interest in encouraging the development and efficient functioning of international markets in which this transfer can occur. In addition, while direct export promotion activities are clearly the responsibility of the Department of Commerce and other federal trade promotion agencies, EPA's international technical assistance and training programs enhance the demand for environmental technologies and services, thereby creating potential commercial opportunities for U.S. business and industry.

The Trade Promotion Coordinating Committee (TPCC) is a interagency group which is sanctioned by legislation, and headed by the Secretary of Commerce, with a mission focused on the harmonization of federal efforts in the export area. Recognizing the broad need for a effective national strategy, the Clinton Administration recently released the report of the TPCC, *Toward a National Export Strategy*. This report established the framework for a unprecedented strengthening of the U.S. government's export promotion efforts. Environmental technologies are the first sector application of this National Export Strategy, and the TPCC kicked off this effort with the publication of *Environmental Technology Exports: Strategic Framework for U.S. Leadership*.

This report, released in November 1993, identifies four specific goals integral to the overall objective of increasing exports of U.S. environmental technologies. They are: to make it easier for businesses to work in partnership with the U.S. government; to strengthen the technology development and commercialization abilities of the domestic industry; to help U.S. businesses succeed in the most important markets today, while developing tomorrow's most potentially valuable markets; and, to coordinate and better focus U.S. governmental export programs and resources.

Responses to Questions From Senator Wallop

Question 4. What is the "Environmental Technology Initiative"? What is its statutory basis? What agencies are involved? What are its FY 94 and proposed FY 95 funding levels?

Answer. In his State of the Union speech on February 17, 1993, President Clinton outlined a new initiative to improve environmental quality and strengthen the American economy. The goal of the President's Environmental Technology Initiative (ETT) is to spur the development and use of more advanced environmental systems and treatment techniques that can yield domestic environmental benefits and increase exports of American technologies to other countries.

EPA's current thinking on how best to implement the President's ETI is contained in our draft Technology Innovation Strategy.* We believe that by fostering the development and use of innovative environmental control technologies, we will be advancing environmental protection and promoting a strong and competitive environmental technology industry.

The draft strategy has four objectives focusing on the functional areas where government intervention can address market barriers facing the industry:

- adapt EPA's policy, regulatory, and compliance framework to promote innovation;
- strengthen the capacity of technology developers and users to succeed in environmental technology innovation;
- strategically invest EPA funds in the development and commercialization of promising new technologies; and
- accelerate the diffusion of innovative technologies at home and abroad.

The strategy has five basic operating principles:

- maximum consultation with stakeholder;
- coordination with Federal, State, and Local agencies;
- partnership and collaboration with the private sector and academia;
- cleaner technology, not just end-of-pipe control technology; and
- measuring progress along the way.

The ETI was developed using existing statutory authorities: the Clean Air Act, the Clean Water Act, the Safe Drinking Water Act, the Pollution Prevention Act, etc. In addition, EPA and the Administration have been working closely with the Committees sponsoring S. 978 and H.R. 3870, not because the Agency lacks legislative authority in this area, but because the proposed bills would greatly clarify EPA's roles in: a) developing public-private technology development and commercialization partnerships; b) extending technical assistance; and c) verifying technology performance.

The FY 94 and proposed FY 95 funding levels are $36 and $80 million, respectively. These amounts are in addition to resources devoted to continuing Agency environmental technology research and development activities.

Question 5. Why is the Initiative important to EPA or DOE activities in the Pacific Rim?

Answer. EPA programs in the Pacific Rim have been extremely influential as these countries develop their environmental management institutions, legislative

---

* Retained in committee files.

infrastructure, and monitoring and research capabilities. However, less has been done in terms of linking these programs to private sector initiatives which are critical to the success of pollution prevention and control strategies. Facilitating the partnering of Asia and U.S. private sectors is essential to ensure the protection of the environment, both in the Pacific Rim and globally.

Question 6. How will the Initiative help American companies to break into Pacific Rim and China energy and environmental technology markets?

Answer. One of the major impediments to the diffusion of U.S. energy and environmental technologies in the Pacific Rim and China markets is a lack of information on U.S. technologies combined with limited opportunity for Pacific Rim decision makers to see U.S. technologies in operation. Through the ETI and U.S. Technology for International Environmental Solutions (TIES) demonstration programs, important problem-solving environmental technologies will be demonstrated. For example, in China, we are planning a demonstration of advanced, but appropriate, flue gas cleaning technology under U.S. TIES. The project will initially evaluate potential sites and the particular technology most suited to the conditions, then demonstrate the technology, opening the door to a vast market for the technology and providing a potential solution to China's largest air pollution problem.

Question 7. Given China's priority on economic development, what importance does China place on environmental protection?

Answer. Same as answer to Senator Johnston's question #3.

Question 14. What activities are EPA and DOE doing in China under the Framework Convention on Climate Change?

Answer. EPA, with DOE, DOS and other federal agencies, is currently implementing the U.S. Initiative on Joint Implementation (USIJI), a pilot program under the President's Climate Change Action Plan. As the program is implemented, we will point towards facilitating proposals for JI projects in China and other countries with significant opportunities for large reductions in greenhouse gas emissions.

In addition, a Chinese proposal for a Climate Change Country Study is currently under review in the second round of the U.S. Country Studies Program. This study will support development of an emissions inventory and help to identify mitigation opportunities in China.

Question 15(a). What is China's position on joint implementation under the United Nations Framework Convention on Climate Change?

Answer: China has not taken a definitive position on joint implementation under the FCCC. They appear to be expressing a cautious interest in the concept of joint implementation while waiting for further international action on the issue.

Question 15(b). Doesn't China endorse the position of the G-77 in opposition to joint implementation?

Answer. The G-77 position is by no means uniform. The statement by Algeria on behalf of the G-77 at the February meeting of the Intergovernmental Negotiating Committee was ambiguous. We believe that this reflects both the diversity of views among developing countries and a general caution due to unfamiliarity with the JI concept. China has not endorsed these statements.

Question 15(c). What is the Administration doing to promote joint implementation projects in China? What arrangements are being made regarding any greenhouse gas reduction credits?

Answer. As previously mentioned, the EPA, with DOE, DOS and other federal agencies, is currently implementing the U.S. Initiative on Joint Implementation (USIJI), a pilot program under the President's Climate Change Action Plan. As the program is implemented, we will point towards facilitating proposals for JI projects in China and other countries with significant opportunities for large reductions in greenhouse gas emissions.

At present the Administration is neither claiming "credit" under the FCCC for Joint implementation actions nor granting credits to private sector actors toward domestic reductions under the climate change action plan. We expect to examine the issue of "credits" in the context of the USIJI. Furthermore, we expect the international community, in its ongoing discussions on this issue in the International Negotiating Committee—and subsequently in the FCCC conference of the Parties—to agree on the appropriate international treatment of "credits."

Question 16. What is China's position regarding the development of a Climate Action Plan for itself? Isn't it opposed to binding obligations for developing countries.

Answer. China submitted a proposal to participate in U.S. Support for Country Studies to Address Climate Change in early 1994. U.S. Support for Country Studies, a two-year $25 million activity, is designed to provide financial and technical assistance to developing and transition countries in support of the Framework Convention on Climate Change. The $1.7 million China proposal requested assistance in four technical areas, including development of national plans and reports in response to the Framework Convention on Climate Change. Specifically, the Chinese have asked for U.S. technical and financial assistance to (1) organize a multi-Ministry climate change policy coordinating body under the auspices of the State Science and Technology Committee (SSTC), (2) identify viable GHG mitigation options in energy and land-use sectors, and (3) organize multi-Ministry workshops in 1995-96 to support the development of a comprehensive national action plan to cope with climate change. A U.S. (EPA, DOE and DOC) delegation visited Beijing in March, 1994, to work with the State Science and Technology Commission to refine the multi-ministry proposal for assistance. The U.S. delegation, after consultation with the U.S. Embassy Science Office, recommended the proposal be supported. Funding decisions regarding support to China will be made following a interagency technical and programmatic review in FY 94. China has not categorically opposed developing country commitments. In fact, by ratifying the FCCC, China has committed itself to "formulate, implement, publish and regularly update national and regional programs containing measures to mitigate climate change. . . ."

Question 28. What activities is EPA conducting in China? How do they relate to DOE's activities? What mechanism is being used to coordinate these efforts?

Answer. Same as answer to Senator Johnston's question #9.

Question 29. Wouldn't the activities being conducted by EPA be more appropriately conducted by DOE? If not, why not?

Answer. Same as answer to Senator Johnston's question #12.

Question 30. What action is EPA taking to coordinate its activities in China and the Pacific Rim with the activities of other Federal agencies?

Answer. All of EPA's activities outside of the U.S. are conducted in accordance with and in coordination with directives of the Department of State. In particular, because the US-PRC Environmental Protocol is governed by the US-PRC Science and Technology Agreement, the Department of State plays an active coordination role in our China program. EPA's Office of International Activities maintains close communication with State and with sister technical agencies regarding ongoing and potential project activities. In addition, we work closely with USAID and other agencies participating in the US-Asia Environmental Partnership on activities in other Pacific Rim nations. EPA also participates actively in the Trade Promotion Coordinating Committee (TPCC) which serves as a focal point for all U.S. government activities relating to technology transfer and export assistance.

Question 31. What is the policy of EPA regarding the use of coal by developing countries? For example, what is EPA's position regarding the use of clean coal technologies by China?

Answer. To the extent that countries rely on coal to meet their energy needs, EPA believes that this coal should be combusted in as efficient and environmentally sound a manner as possible. Clean coal technologies are clearly an important element of meeting this goal. In addition to clean coal technologies, EPA also encourages countries to develop other environmentally beneficial energy sources, including natural gas and renewables and to use energy as efficiently as possible in all sectors of the economy.

Question 32. Why should EPA take a lead in environmental technology development when it is a regulatory agency? Doesn't this create a conflict of interest?

Answer. There are several reasons why EPA should take a lead in environmental technology development. First, one of the major ways in which technology development can be encouraged is to adapt EPA's policy, regulatory, and compliance framework to promote innovation. This can only be addressed by EPA. In addition, some programs can only be run by EPA, such as regulatory approvals of new environmental monitoring methods. Second, new environmental technologies often face difficulty in penetrating the marketplace. EPA verification of the performance of environmental technologies is sought by companies to provide a official imprimatur needed to market their technologies. Third, although EPA remains primarily a regulatory agency, we are increasingly recognizing that to succeed in our environmental protection goals, we must adopt a variety of non-traditional approaches, be they promoting needed technological innovation, pollution prevention, or relying on voluntary and other cooperative programs to address unregulated sources or otherwise intractable problems.

From EPA's perspective, promoting environmental technology is a complementary program, necessary to our basic mission, not a conflict of interest. EPA sees promotion of technology innovation as a necessary tool to help achieve our environmental goals. As such, its promotion is constrained not to conflict with the ends we are seeking.

Taking a leading role nevertheless requires coordination with other federal agencies so as to more effectively utilize limited federal resources. EPA is working with the Departments of Commerce and Energy, the Agency for International Development, the Small Business Administration, the Export-Import Bank, and the Overseas Private Investment Corporation to implement the Administration's Environmental Technology Export Strategy. In addition, EPA's FY 95 ETI planning efforts involve representatives from other federal agencies.

Question 33. In your testimony, you mention that "one of the impediments to U.S. efforts to reduce China's greenhouse gas emissions enough to stabilize atmospheric concentrations is limited foreign dollars from all countries."

a. While not required by the network Convention, do you mean to suggest that the U.S. or other developed countries, or both, have an obligation to reduce China's greenhouse gas emissions enough to stabilize atmospheric concentration?

b. Do you mean to suggest that all countries other than China have an obligation to subsidize with foreign dollars the efforts of China to reduce their greenhouse gas emissions?

Answer. Under the Framework Convention Annex II Parties (including the U.S.) are committed to providing Financial resources to help developing country Parties (including China) meet their commitments under the FCCC, and to facilitating the transfer of environmentally sound technologies to those Parties. China currently has no specific emissions reduction aim under the Convention.

The Administration would like China and other developing countries to establish national emissions inventories and commit to emissions reductions as soon as possible. The U.S. Country Studies Program was established to help developing countries build capacity to undertake such efforts.

APPENDIX II

Additional Material Submitted for the Record

CONGRESSIONAL VISIT TO CHINA, VIETNAM, THAILAND AND INDONESIA

INTRODUCTION

In January 1994, a bipartisan delegation of United States Senators visited China, Vietnam, Thailand and Indonesia to assess first hand the issues in the relationships of the United States with these key Asian nations. These issues are of immediate importance to Congress and the President. They deserve the sober consideration and close study that personal experience with the individual countries and their leaders can provide far more efficiently than a thousand printed documents. The stakes for the United States in Asia are enormous. Congress will debate the future of these relationships later this year, for example, when the question of most favored nation trade status for China is considered.

In each country the delegation had the privilege of meeting personally with officials at the highest level and engaging in unconstrained discussion of a broad range of subjects, including trade, foreign policy, human rights, religious freedom, national security, economic policy, energy supply and use and science and technology. This report presents the findings and recommendations of the majority of the delegation and includes an appendix containing background information gathered on each of the countries visited.

Senator Nickles does not associate himself with the views expressed in this report.

The following Senators comprised the delegation:

Senator J. Bennett Johnston, delegation leader
Senator Mark O. Hatfield (China, Vietnam and Thailand)
Senator Alan K. Simpson
Senator Arlen Specter
Senator Don Nickles (China, Vietnam and Thailand)
Senator Harlan Mathews
Senator Robert Bennett

I. CHINA

Visits

In China, the delegation visited both the capital city of Beijing and Guangzhou, the center of the great economic boom in South China. Upon arrival in Beijing, the delegation received a briefing from Ambassador Stapleton Roy and his team at the U.S. Embassy. The next day was spent meeting with officials of the government of the Peoples Republic of China, including Premier Li Peng, Mr. Qiao Shi, the chairman of the Standing Committee of the National Peoples Congress (NPC) and Mr. Zhu Liang, chairman of the Foreign Affairs Committee of the NPC. Government ministers visited included Mme. Wu Yi, Minister of Foreign Trade and Economic Cooperation, Vice Foreign Minister Liu Huaqiu, and Mr. Huang Quitao, Vice Minister of the State Science and Technology Commission, who chaired a briefing by several ministries involved in the development and implementation of Chinese policy for energy, science and technology.

In Guangzhou, the delegation met with U.S. Consul Eugene Martin and his staff and with local Chinese officials, including Mayor Li Ziliu, Mr. Zhang Gao Li, Vice Governor of Guangdong Province, and Mr. Yu Fu Min of the Guangdong Provincial Electric Power Bureau. The delegation also visited private residences in Guangzhou. Senator Hatfield and Senator Nickles met privately in Guangzhou with the Rev. Samuel Lamb to discuss religious freedom in China.

At each stop the delegation had breakfast with members of the local chapter of the American Chamber of Commerce.

Findings

1. Importance of the relationship

China now has one of the world's largest economies. Many experts predict that China's economy will become the world's largest in the next century. In the last two years the inflation-adjusted growth rate in China has been 13 percent. Over the last 15 years China's growth rate has averaged 9 percent per year.

The enormity of the Chinese market holds lucrative trading potential for the United States. Nowhere is this more apparent than in the growing demand for new electric power capacity in China. Currently, the electrical generating capacity for China's 1.2 billion people is only 165 gigawatts (Gwt). This compares with roughly 800 Gwt for the 250 million people of the United States.

To keep up with this explosive demand for electricity, China's 30 provinces plan to add 15 Gwt per year through the year 2000 and still larger annual amounts in the years between 2000 and 2010. Just four provinces in southern China expect to build the equivalent of the entire Bonneville Power Administration system and do it by the end of the decade. These same four provinces have 20 new airport projects underway, one of which is the size of Chicago's O'Hare International Airport.

These infrastructure projects can provide long-term markets for the United States, but only if the U.S. government allows U.S. firms to compete effectively with those of other nations. The rapidly developing Chinese middle class represents the most promising growth market in the world for consumer products. Yet current U.S. policies do not encourage U.S. companies to enter this market, and in several instances, these policies actively discourage investment in China.

The long relationship between the United States and China has a rich history that began with the American clipper ships that helped reestablish contact between China and the rest of the world in the 19th century. This relationship will be one of the most important for each nation in the future. There is a solid foundation for mutual respect and cooperation based on critical periods in history when both nations have worked closely together.

Today both nations have key economic contributions to make to a global world market that is increasingly integrated in structure. The United States will continue to play an aggressive leadership role in promoting international economic cooperation. Recent foreign policy achievements of the Clinton Administration, such as the first leaders' meeting held in conjunction with the Asia Pacific Economic Cooperation (APEC) organization's annual ministerial meeting in Seattle, the approval of a new revised General Agreement on Tariffs and Trade (GATT) and the enactment of the North American Free Trade Agreement (NAFTA), will accelerate the reduction of trade barriers around the world. The U.S. and China need to work together to take advantage of this momentum and eliminate barriers to economic cooperation that limit the economic integration of the United States with China and other Pacific Rim nations.

The U.S. and China must also cooperate if both nations are to play an effective role in fostering continued peace and mutual security in Asia and the Pacific. There is an urgent need to limit the proliferation of nuclear weapons and to act to defuse issues in dispute now being addressed through military competition or the threat of armed conflict. Here the U.S. and China should lead by cooperation and dialogue and setting forth strong examples of mature and restrained behavior.

2. Economic progress and human rights

It is especially important this year that the U.S. and China work together constructively to emphasize positive aspects and defuse potentially disruptive elements of the relationship.

There are many positive common characteristics that provide a basis for strengthening the relationship. China, like the United States, is a large, geographically diverse country with considerable indigenous natural resources and a hard-working population. The people of both countries put a premium on education and welcome new ideas and advanced technologies. Both nations prefer regional and private enterprise to centrally run economic institutions, and in this they differ from most of the nations of the world. Both foster local autonomy and self reliance. Both encourage individuals to work to better their own situation.

The economies of China and the U.S. complement each other. China produces high quality labor intensive goods for domestic consumption and export. The U.S. is an important customer for these exports, but is far from the only such customer. The U.S. emphasizes the high technology and managerial skill the Chinese economy needs.

Economic growth continues to improve the quality of life and choices available to individual Chinese. The 1990's are the best years any modern generation of Chinese have seen, and the desire to continue to improve the quality of life of the average Chinese citizen can be seen at all levels. Involvement of U.S. business in China has contributed significantly and positively to this progress by introducing new management methods that emphasize individual choice and by setting an example of practice that inspires respect and imitation.

However, the U.S. continues to be concerned with the human rights performance of the Chinese government. These concerns, expressed by President Clinton on behalf of a vast majority of Americans, reflect a reality of U.S. internal politics that cannot be ignored. Any hope of reaching a stable framework for economic cooperation between the two countries that is not threatened politically on an annual basis depends on an improving human rights record. China must realize that, for better or worse, the Chinese government's treatment of its citizens will be publicized worldwide and will have political consequences in democratic countries like the United States.

The U.S. will continue to emphasize to the Chinese government its interest in progress in human rights and religious freedom for individuals. The Chinese for their part will continue to emphasize their view that human rights include rights to basic nutritional subsistence, personal security, economic development, educational and economic opportunity and domestic stability. Dialogue, negotiation and mutual respect for the perspectives each nation brings to this debate can strengthen the relationship. Strengthening the economic ties between our two nations is the surest path to progress on human rights for 1.2 billion Chinese. Trade sanctions and unilateral disengagement by the U.S. would be unnecessarily destructive of the relationship and would almost surely reduce the pace of progress in human rights in China if power is shifted to old-line reactionary Communist elements in the Chinese political system who favor autarky and oppose most things Western.

Members of the delegation had an opportunity to meet with Rev. Samuel Lamb, a prominent leader of one of the major house churches that operate outside the officially sanctioned churches of China: The members were impressed with the determination of these Chinese Christians in their untiring pursuit of personal exercise of religious freedom. Although the constitution of China officially guarantees religious freedom, there have been many periods when official opposition to house churches became active religious persecution. However, in the experience of members of this church visited in Guangzhou by delegation members, government pressure on unrecognized churches lessens when relations between China and the U.S. improve.

3. Mutual security, weapons sales and nuclear proliferation

An Asian arms race in general and the spread of nuclear weapons in particular work against the security interests of both the U.S. and China. Each country sells weapons in Asia. Each country should avoid weapons sales that are destabilizing and provocative. In particular, China needs to respond forthrightly to allegations by some in the U.S. intelligence community about illegal missile sales to Pakistan. If the President finds illegal sales have occurred the operation of current law require actions that could severely damage the U.S.-China relationship. It is important that this issue be clarified completely so it can be dealt with and put behind us.

President Clinton has clearly stated the absolute opposition of the U.S. to any plans by North Korea to develop nuclear weapons. China, which shares a border with North Korea, has said often that it favors a non-nuclear Korean peninsula. The delegation was told that China is willing to work quietly with the U.S. to eliminate the possibility of a nuclear-armed North Korea. As major powers in the Asia-Pacific region, the United States and China need a unified and coherent approach to the problem of nuclear proliferation on the Asian mainland. We do not have such an approach now. China should work with the United States to devise a joint approach to this most important issue.

4. Future developments

Asia is fast becoming the most important trading region in the world for the U.S., and China should become our most important trading partner. Premier Li Peng estimated that China would import $1 trillion in goods, services and investment between now and 2000. The United States has a great responsibility to help the historic modernization of China proceed rapidly and in an atmosphere of security and stability. The United States and China are great nations who have many common characteristics and interests and should develop common approaches to as many issues as possible.

Recommendations

1. PRESIDENT CLINTON SHOULD DETERMINE THAT HUMAN RIGHTS CONDITIONS HAVE IMPROVED IN CHINA TO AN EXTENT SUFFICIENT TO

EXTEND MOST FAVORED NATION (MFN) NONDISCRIMINATORY TRADE STATUS WITHOUT CONDITIONS AND SHOULD PROPOSE LEGISLATION TO MAKE UNCONDITIONAL MFN FOR CHINA PERMANENT.

The majority of the delegation believes that the best way to achieve progress in human rights in China is by continuing to engage China politically and by strengthening and accelerating involvement of the U.S. in the development of the Chinese economy. To ensure continued progress and improve the ability of the U.S. to influence China, the U.S. must find a way to advance its human rights concerns through some mechanism other than the annual debate under current law on most favored nation (MFN) trade status for China. Threatening to deny MFN status is an exceedingly blunt instrument that offers policy makers an all-or-nothing choice with none of the flexibility needed to advance the diversified goals the U.S. seeks in its relationship with China. Because of the great stakes in the U.S.-China relationship threats to deny MFN status seem extreme and at the same time lacking in credibility.

The delegation noted progress in many areas enumerated in President Clinton's Executive Order of May 1993 on human rights and MFN status and anticipates that further progress will be made before the President must make determinations as set forth in the Executive Order. Specifically, the delegation is aware of:

(a) Discussions underway with the International Committee of the Red Cross to allow access to prisons and prisoners to ensure humane treatment;

(b) Cooperation with U.S. Embassy and U.S. Customs officials to resolve concerns about alleged use of prison labor to produce goods for export to the U.S.;

(c) Release of some Chinese citizens imprisoned or detained for nonviolent expression of their political or religious beliefs, including Democracy Wall and Tiananmen incident activities; and

(d) Bilateral discussions between representatives of the Dalai Lama and the Chinese government on the religious and cultural concerns of the people of Tibet.

The delegation urged many Chinese officials in meetings to work to accelerate progress in these and other areas of concern to allow President Clinton to find that overall significant progress on human rights has occurred as required by the Executive Order. The delegation was assured by Chinese officials, including Premier Li Peng, that human rights conditions would improve faster and in a more lasting way through bilateral dialogue and discussion. Linking trade and human rights through MFN does not appear necessary to ensure progress in human rights and in the future could limit such progress if this linkage results in a severe crisis in the U.S.-China relationship.

The delegation also noted the great improvement in the ability of the Chinese to meet basic human needs as a result of economic development and reform in China, which is also increasing access to outside sources of information. Economic development has resulted in greater individual choice in residence, education, employment, and movement throughout the country. Stable domestic conditions have aided these improvements.

The delegation was also told that important new initiatives are underway to establish a system of rule of law, beginning with a legal foundation for the economic changes that are occurring in China. Such a system is essential for the success of the market reforms the central government has proposed and has the potential over time to enable individuals to use legal means to exercise their rights. The majority of the delegation believes it is important for the U.S. to contribute technical and managerial assistance to this effort.

2. REINSTATE OVERSEAS PRIVATE INVESTMENT CORPORATION (OPIC) COVERAGE FOR U.S. BUSINESSES IN CHINA; REESTABLISH ELIGIBILITY OF CHINA FOR TRADE AND DEVELOPMENT AGENCY (TDA) ASSISTANCE PROGRAMS AND RELAX CURRENT EXPORT-IMPORT BANK RESTRICTIONS APPLYING TO CHINA.

Under current regulations there is a $50 million cap on Export Import Bank loans and guarantees unless a presidential waiver is granted. This cap and waiver requirement should be eliminated. The U.S. should consider allowing nonrecourse loans to project entities and local governments. U.S. financial assistance should encompass project financing.

Since 1989, a number of bilateral aid programs have been prohibited from operating in China, first by administrative action and later by statute (P.L. 101-246), in an attempt to put pressure on the central authorities to respect internationally recognized human rights. However, some of these prohibitions eliminate benefits to U.S. business, not the central government. Others primarily affect benefits to regional authorities and ordinary citizens.

The majority of the delegation believes the President should review the restrictions on OPIC, TDA, and the U.S./Asia Environmental Partnership and, in particular, on Export Import Bank programs to determine whether the requirements for lifting these prohibitions and restrictions have been met. More generally, the president's review should consider whether these restrictions are achieving the results they were instituted to achieve.

3. LIBERALIZE LICENSING PROCEDURES FOR U.S. EXPORTS CONSISTENT WITH PRACTICES ADOPTED BY OUR MAJOR COMPETITORS IN THIS AREA.

At the end of March 1994 the current multilateral export control system for high technology exports known as the Coordinating Committee (COCOM) regime is scheduled to expire. As the Administration develops recommendations for a replacement regime, consideration should be given to changing the presumption on license applications to one of approval unless specific findings are made within a reasonable time certain. Additional useful proposals, outlined in the Department of Commerce's September 1993 report entitled *Toward a National Export Strategy,* would streamline and simplify U.S. export control regulatory requirements. In general, U.S. policy should be aimed at creating a competitive environment for U.S. firms seeking to do business in China relative to competitors from other nations active there.

4. ENCOURAGE A HIGHER DEGREE OF COOPERATION IN TRANSFER OF HIGH TECHNOLOGY AND MANAGERIAL EXPERTISE: EXPLORE OPPORTUNITIES FOR ENHANCED COOPERATION BETWEEN THE DEPARTMENT OF ENERGY LABORATORIES AND APPROPRIATE ENTITIES IN CHINA, PARTICULARLY IN THE AREA OF ENVIRONMENTAL TECHNOLOGIES, SUCH AS ENVIRONMENTAL REMEDIATION AND WASTE MANAGEMENT, ENVIRONMENTALLY CONSCIOUS TECHNIQUES AND PRACTICES AND RENEWABLE ENERGY AND ENERGY EFFICIENCY TECHNOLOGY AND PRODUCTS.

Several Chinese officials expressed a desire for increased access to U.S. technology, which the Chinese believe is in many instances superior to the technology offered by our competitors. The Chinese are also anxious to learn about the management techniques used by the U.S. private sector, which they regard as the most proficient in the world. This interest should provide many opportunities for cooperation, which can in turn accelerate the voluntary adoption of the practices that U.S. policy is urging on China.

Chinese officials expressed concern about the environmental consequences of the rapid development that is taking place, but seem to regard these consequences as unavoidable for a relatively poor nation. The U.S. has learned through experience that improved technology can in fact mitigate many of the adverse consequences of development. There is a tremendous opportunity to apply efficient, environmentally-conscious technology and practices developed and tested in the U.S. to assist China's development without the environmental negatives experienced in the industrialization of the West. U.S. policy should encourage this.

The Department of Commerce made a number of findings in the report *Toward a National Export Strategy* that if implemented would facilitate environmental technology exports. The report called the environmental technology export market "one of the most attractive arenas of U.S. Government focus to enhance our international competitiveness and export performance." Worldwide, spending on environmental technology in 1992 ranged from $350 to $380 billion, and could be as high as $590 billion by the year 2000. Pacific Rim countries will double spending in this area, particularly for low technology exports in solid waste management, water delivery and treatment and air pollution control.

Like our major competitors, the U.S. Government has programs to develop and assess this market, but as the Commerce Department report found, U.S. programs are substantially less aggressive and less comprehensive than those of our major competitors. The Department of Energy, for example, leads the Committee on Renewable Energy Commerce and Trade (CORECT), a public-private working group to promote U.S. energy products and services internationally and also holds trade missions and conferences to promote exports in the energy sector. DOE has been woefully lacking in its efforts in China in these activities. Another major DOE program, the Federal International Energy Trade and Development Opportunities Program (FIETOP) is described by this same Commerce Department report as "inactive". Other DOE-sponsored feasibility study programs involving areas such as the promotion of Clean Coal Technology do not as yet fully operate in China, although there appears to be no statutory obstacle to such operations.

The Administration and the Congress should review these Department of Energy programs, and those of other agencies, to determine which would help meet the needs of China and other East Asian countries in developing a plan to coordinate energy and environmental needs. Resources should be reallocated to assure the implementation of an aggressive technical assistance and training program for China and East Asia as soon as possible.

Similarly, the Administration and Congress should create a lead laboratory among the Department of Energy Laboratories for developing a comprehensive program for providing technical assistance and training for China and other Asian countries in the areas of environmental technologies and control.

5. REVISE U.S. RULES REGARDING SALE OF U.S. NUCLEAR POWER TECHNOLOGY IN CHINA, SINCE CHINA HAS SIGNED THE NUCLEAR NON-PROLIFERATION TREATY.

China has signed the Nuclear Non-Proliferation Treaty and enforces a policy for nuclear exports that is in compliance with the NPT. A positive determination of human rights progress for purposes of extension of MFN would also clear the way under U.S. law for the export to China by U.S. firms of nuclear power technology. China has plans for significant nuclear power development: 6 gigawatts are expected to be on line by 2000, an additional 6 gigawatts will be under construction at that time and acceptable experience could lead to rapid expansion in nuclear generation thereafter. The Chinese are firm in their belief that nuclear power must be a part of the solution to growing air quality problems throughout China from over-reliance on coal. China is considering French, Japanese and even Russian nuclear power technologies to meet this need. U.S. technology has not only obvious commercial advantages for our economy but clear safety advantages for the Chinese. The majority of the delegation can find no reason why the U.S. continues to restrict U.S. firms' entry into competition with other nations for the construction of nuclear plants in China that will take place in any event.

The U.S. government should seek to identify mechanisms for dialogue and discussion that offer constructive possibilities for resolving U.S. concerns about nuclear weapons proliferation. The current practice of blanket denial of permission for U.S. nuclear suppliers to export to China is counterproductive and inflexible. There is considerable respect for the quality and performance of U.S. civilian nuclear technology among knowledgeable Chinese. A creative export policy for nuclear technology could use this advantage to further U.S. proliferation, nuclear safety and environmental goals simultaneously by allowing U.S. nuclear firms to compete for business in China.

Nuclear power technology is one of the most politically sensitive examples of a dual-use technology that has significant trade potential in the Pacific Rim. A dual-use technology is a technology that has simultaneous utility, directly or indirectly, in civilian and military applications. The United States needs to develop an export policy with respect to dual-use technologies that appropriately balances the trade advantages of allowing market forces to operate with the national security and weapons proliferation considerations inherent in such trade.

In the case of nuclear power, the NPT provides a framework for judging the security and proliferation impacts of trade in specific nuclear equipment. China has signed the NPT and therefore deserves the same treatment under policies for export of nuclear power technology that is accorded other signatories of the Treaty. A clear U.S. policy for exports of dual-use technology is needed for nuclear technology, but also for other technologies, such as computers and certain computer software, rockets, satellites, communications and surveillance, and other high-technology applications of research and development that have dual application to civilian and military missions.

The U.S. needs to address this issue. Sales of dual use technologies do occur and are accelerating worldwide no matter what the U.S. decides to do, because these technologies, by definition, offer real value to the civilian sector. The U.S. holds no permanent monopoly in any of these technologies. The actions of the French, Japanese and Russians in seeking to satisfy the Chinese demand for nuclear power technology provide an obvious example of the inability of the U.S. to control such trade by unilateral action. An intelligent export policy would recognize that trade in dual-use technologies is occurring and would seek to balance the goals of enhancing our own economy, increasing U.S. leverage over the behavior of customers for dual-use technology and enhancing security worldwide. The U.S. should the process of developing such a policy.

Senator Hatfield's views on this matter differ from those expressed in this recommendation.

6. ENCOURAGE U.S. GOVERNMENT CONSULTATIVE MISSIONS TO COORDINATE THEIR ACTIVITIES WITH COMMERCIAL ADVISORY GROUPS

TO MAXIMIZE OPPORTUNITIES FOR U.S. BUSINESS TO INCREASE MARKET SHARE IN CHINA, NOW THAT SUCH CONTACTS ARE BEING RESUMED.

The delegation believes that support for U.S. business in China by the U.S. government is an important component of U.S. policy in China. Government and business should work together. U.S. businesses in China are an important resource for understanding developments in China and in demonstrating the value of U.S. ideas, such as respect for individual rights.

As an example, the delegation heard from one American businessman in Guangzhou who described his firm's successful effort to introduce Chinese universities to the concept of job recruitment of university graduates. Practice only a few years ago was for central authorities to assign all university graduates to a job. Today many Chinese students can look forward to bidding by multinational companies for their services as a result of these efforts by private business. This is one small example of how the involvement of American business is increasing individual choice and breaking down the centralized control that has characterized Chinese life until recently.

7. THE U.S. MUST FUNDAMENTALLY ALTER ITS APPROACH TO CHINA (AND OTHER LARGE DEVELOPING NATIONS IN ASIA) IF EFFORTS AT COOPERATION ON GLOBAL ENVIRONMENTAL ISSUES ARE TO HAVE ANY CHANCE OF SUCCESS.

The cost to the environment of the unparalleled industrial growth and urbanization occurring in China and throughout Southeast Asia was obvious to every member of the delegation. In Beijing, Bangkok and Jakarta the delegation experienced firsthand the problems every resident has daily with air pollution. In Guangzhou, Jakarta, Hanoi and Ho Chi Minh City, water pollution problems were evident. Lack of adequate solid waste disposal and primitive or nonexistent waste water treatment afflicted virtually every major urban area. Deforestation and subsequent impacts on soil erosion and water quality were apparent in many parts of China, Vietnam, Thailand and Indonesia.

One of the most serious problems facing Asia is growing and uncontrolled particulate and sulphur dioxide emissions. The World Bank has reported that in Bangkok, Jakarta and China's larger cities including Beijing, ambient levels of particulate matter exceeded World Health Organization (WHO) standards 100 days or more each year. This is worse by a factor of ten compared to the situation in the most developed countries.

Similar growth exists in CO_2 emissions, caused largely by the burning of fossil fuels, principally coal. The World Bank estimates that a 20 percent reduction in annual carbon dioxide emissions in all industrial countries will be offset by emissions in East Asia within ten years. Some estimate that China alone will overtake any gains made by industrial countries in reducing CO_2 emissions if China's current growth rate is sustained.

There are profound global environmental implications in these trends. If left unabated, actions industrial countries take to limit total levels of emissions within their own borders will be swamped by growth in emissions in Asia, and overall levels of greenhouse gases and acid rain will accelerate rather than stabilize or decrease. The delegation saw strong evidence nearly everywhere that the impact of growth on the environment in Asia threatens the ability of the nations of the world to meet the goals set forth in the Rio Conference on the Environment in 1992.

There are serious health implications for local populations in Asia unless major efforts are undertaken soon to address these problems in a comprehensive way. The linkages between particulate emissions and respiratory disease are well documented; linkages with lead poisoning and associated mental dysfunctions in children have more recently been associated with air pollution and are increasingly a health concern.

Water pollution is the most widespread environmental problem in Asia. The delegation saw particular evidence of this in the Pearl River watershed below Guangzhou. The pollution of the Pearl River has implications beyond China's borders since both Hong Kong and Macau depend on this watershed for water supplies. The delegation was told, for example, that only six of Guangdong's largest twenty cities have waste water treatment plants. Only ten percent of residential waste water in Guangdong Province is treated and that ten percent only receives primary treatment. As the World Bank and others have noted, water pollution has a severe impact on the poor and among children, and has been well documented as a major cause of mortality and illness.

United States policies towards trade with China dating from the Tiananmen Square incident in 1989 severely limit the ability of U.S. institutions to provide technical and other assistance to China to plan and implement programs to control and lessen environmental degradation. These limitations are prolonging widespread pollution from inadequate solid and liquid waste disposal and deforestation. The U.S./Asia Environmental Partnership is prohibited from operating in China, even though air and water pollution can never be substantially reduced in Asia without coordinated strategies which include China. The Trade and Development Assistance (TDA) program, which funds feasibility studies, orientation visits, training grants and various technical assistance programs, also is prohibited from operating in China. Ironically, these studies would benefit precisely the people the U.S. government claims to be trying to help through the post-Tiananmen sanctions: the men, women and children of China who aspire for a better life.

Under these assistance programs, the funds and benefits do not flow to the central government in Beijing that the U.S. holds responsible for the Tiananmen Square incident, but to regional authorities who will necessarily implement environmental protection and remediation projects and programs. In South China, for example, where water pollution is a problem of crisis proportions, some 80 percent of all infrastructure projects are funded through the regional government, not the central government. And in Beijing, it is the municipal not central government authorities who are taking the lead in seeking help in dealing with serious air quality issues.

It is time for these restraints to be lifted, not only for humanitarian reasons, but also because it is in the U.S. interest to do so. The major share of the technologies that will be deployed to address these issues could be-US. technologies if the political barriers were cleared away and U.S. companies allowed to compete to perform environmental work in Asia.

The needs are so huge for pollution abatement that financing is beyond the ability of current international institutions and individual donor countries. The private sector must be more deeply involved if the amount of financing required is to be made available. However, self-imposed restrictions on U.S. bilateral programs that could facilitate such participation are severely constraining private sector involvement in environmental and other infrastructure projects. These restrictions include the current prohibition on OPIC guarantees for U.S. firms doing business in China and Vietnam, limits on Export-Import Bank programs and restraints on the use of projected revenue streams from user fees to secure loans.

The majority of the delegation strongly recommends that the Administration undertake an immediate review to identify and remove restraints on U.S. export promotion programs. The Administration should also undertake a review of U.S. credit programs to determine whether new initiatives are needed to help finance initial capital expenditures that will be required to address the environmental problems in China, Vietnam, Thailand, Indonesia and other nations of Asia.

Such a review should also address the critical training needs for planners and regulators in these nations. If environmental problems are to be dealt with successfully, each of the nations the delegation visited must become more self sufficient in developing and implementing environmental planning and management programs, especially at the local and regional levels. The Administration should review current U.S. export program restrictions and determine what revisions are needed to undertake effective training programs for such personnel in these nations, and what changes are needed in current programs to facilitate the transfer of environmental technology. Of particular importance are basic, low cost programs and technologies that are politically and economically feasible in nations that will continue to spend the major share of scarce investment funding on economic development.

II. VIETNAM

Visits

In Vietnam, the delegation visited the capital city of Hanoi and Ho Chi Minh City (formerly Saigon). Upon arrival in Hanoi, the delegation was greeted by Vice Foreign Minister of Vietnam Le Mai and was extensively briefed by Major Gen. Thomas Needham, Commander of the U.S. Joint Task Force on Full Accounting, and his staff. The delegation sponsored a luncheon for members of the Vietnamese National Assembly and met with Vice-Minister of Energy Le Liem and Nguyen Tri Lien, the Vice-Chairman of Petrovietnam. The delegation had a long visit with Do Muoi, the Secretary General of the Vietnamese Communist Party, the top official in Vietnam, and toured the Vietnamese War Museum and Joint Archives Center. The delegation was hosted at lunch by the Vietnam Chamber of Commerce and Industry, a group of Vietnamese private business men and women.

In Ho Chi Minh City the delegation met with the American business community and a delegation of city officials headed by Nguyen Van Huan, the First Vice-Chairman of the Ho Chi Minh City Peoples Committee. The delegation visited the Central

Vocational School for Handicapped Persons, a project outside Ho Chi Minh City that receives support from U.S. AID, and visited the Linh Trung Export Processing Zone nearby.

The delegation was very pleased to be able to offer Major Nguyen Quy An and his daughter passage to Bangkok on the delegation airplane. Major An received the Distinguished Flying Cross for bravery during the Vietnam War and had been trying for over two decades to leave Vietnam.

Findings

1. Timing for improvement of U.S.-Vietnam relations

In the Peoples Republic of China, it is clear that great changes are taking place—within China and in China's relations with the United States. While there are vast differences between China's situation and that of Vietnam, there is a similar opportunity for great changes in Vietnam and in the U.S.-Vietnam relationship. Vietnam has recently adopted an outward-looking policy and is seeking normal relations with the U.S. and all other nations. The Vietnamese economy has been growing at 7 to 8 percent per year, but the base remains that of a severely underdeveloped country.

Recently, the way was cleared for international financial institutions to invest up to $1.9 billion in improving the inadequate Vietnamese infrastructure in transportation, telecommunication and energy supply. Overall, Vietnam hopes to obtain $20 to $25 billion in foreign assistance during the rest of the current decade, more than half of which will come from private investment, in order to achieve its goal of at least 7.5 percent real annual growth in this period. The projects that flow from these investments will determine trade and investment opportunities in Vietnam for many years to come. At present Japanese, French, South Korean, Taiwanese and Australian companies are best positioned to take advantage of the commercial opportunities flowing from new multilateral aid.

There is a good prospect that Vietnam will continue to experience rapid economic development and will soon take its place as a valuable member of integrated Asian and global economies. The United States could be a great help in moving this process forward. The Vietnamese are very desirous of developing closer relations with the U.S. to gain access to U.S. capital, technology and managerial experience. The Vietnamese government officials the delegation met seem genuinely to value American participation in the development of their economy and appear determined to make rapid progress in improving the lives of the Vietnamese people.

The majority of the delegation believes that the time is right for action by the President to improve in U.S.-Vietnam relations. Closer relations will allow participation by U.S. firms in the development of the Vietnamese economy on a scale that will not be available a year from now. There is an opportunity now to greatly increase the influence of American technology, practices and ideas in Vietnam. This opportunity will not only further achievement of long held goals with respect to POW/MIA issues, but could usher in a period of great benefit to both countries and the Asia-Pacific region generally.

2. POW/MIA issues

Because of the complex and in part tragic history of the U.S. relationship with Vietnam, the POW/MIA issues are in the forefront of the internal politics of the United States whenever Vietnam is discussed. These concerns and their impact on U.S. policy are very real and cannot be ignored.

Based on the briefings and other information provided to the delegation it is evident that the Vietnamese government is cooperating as fully as we have a right to expect in the search for missing U.S. servicemen in Southeast Asia. One American official on the ground in Vietnam noted that in his view, the U.S. government could not provide a foreign government this kind of access if the situation was reversed.

Some say the U.S. will better achieve full accountability on all POW/MIA issues by continuation of the embargo on trade with Vietnam to keep pressure on the Vietnamese government. Others say the U.S. will make much more progress if we act soon to support progressive elements in the Vietnamese political system by normalizing our relations with Vietnam. The delegation was assured by U.S. military leaders in Vietnam and repeatedly by the Vietnamese, including direct personal confirmation by Secretary General Do Muoi, that cooperation on POW/MIA issues would continue beyond the end of the embargo. A bipartisan majority of the delegation believes this is the most important reason to lift the embargo and expects cooperation to improve significantly as relations improve and U.S. engagement with Vietnam proceeds. This majority believes that the most effective response of the U.S. government to the cooperation afforded to date to the U.S. Joint Task force team so far would be a permanent lifting of the embargo.

Based on the evidence provided to the delegation, it appears that positive movement on the U.S. side is very important to ensure that the POW/MIA effort continues to make the progress towards full accountability experienced in the last two years. Failure of the U.S. to show good faith in some meaningful fashion within the next few weeks or months could seriously damage the strong efforts undertaken so far. A bipartisan majority of the delegation believes that total elimination of the embargo and prompt movement towards recognition of the Vietnamese government would be the kind of meaningful action required at this time. Senator Nickles and Senator Specter sought further information and consultations before deciding this issue.

3. Economic and political issues

One purpose of the visit by the delegation was to determine whether there is a basis for closer economic ties between the United States and Vietnam. Vietnam is making significant economic progress on its own and in cooperation with Japan, South Korea, Taiwan, China, Australia, France, the UK and other nations. Only the U.S. is missing from the roster of significant investors in the Vietnamese economy.

Vietnam is not yet an Asian "tiger". Per capita GDP in Vietnam is only $200. However, the Vietnamese population is highly literate, eager for advanced technology and motivated to work. The Vietnamese economy is acquiring a reputation for being well run. The Vietnamese government is beginning to recognize, as many countries have found from bitter experience, that centrally-run economies don't work well. Decollectivization of agriculture has resulted in increased rice production to the point that Vietnam now challenges the U.S. and Thailand as a leading world exporter of rice. Currently international financial institutions estimate that 60 percent of Vietnam's GDP comes from its growing private sector. Vietnam has significant plans for further privatization and should consider accelerating this effort.

The Vietnamese government has acted to deal firmly with corruption, which is a significant threat to the trust that market solutions require.

Vietnam, like other developing Asian nations, such as China, Indonesia and Singapore, falls short of U.S. standards of openness in political participation. The majority of the delegation believes that economic liberalization will lead inevitably to political liberalization. No economy dependent on individual initiative and choice will fully succeed unless the rights of those individuals are respected by government. The ability to monitor and influence Vietnamese treatment of their citizens will be significantly enhanced over the current situation by official U.S. presence in Vietnam and establishment of a normal bilateral relationship.

Many of the reform efforts under way and reasonably anticipated will benefit the Vietnamese people, improve individual opportunity, and earn respect for Vietnam. Trade and closer engagement with the U.S. on a broad front will accelerate this process. Continuing to turn our backs on Vietnam will only strengthen the indigenous forces opposing reform in Vietnam and disadvantage U.S. individuals, organizations and firms that hope to and could participate in improving the lives of the Vietnamese people.

4. Future developments

Asia and the Pacific rim are becoming the most important trading regions for the U.S. Vietnam should also become an important U.S. trading partner. The U.S. has an important role to play to assist Vietnam in making the transition to a deserved and full partnership in the community of nations.

In both relationships—with China and with Vietnam—the U.S. needs to find a mechanism to express our interest in democratic values and individual freedom without sacrificing the potent impetus for reform provided by economic, political and cultural engagement with these rapidly developing market economies. The U.S. should not temper its insistence that respect for the individual is fundamental to just governance. However, the U.S. will be most effective in influencing Vietnam and China towards this view if more and more Americans are involved in the economic life of these countries. The articulation of a new relationship with China and Vietnam not driven by MFN status is an essential task for the President over the coming months. This new relationship must combine engagement with China and Vietnam with effective pressure for improvement in human rights.

Recommendations

1. THE PRESIDENT SHOULD LIFT THE EMBARGO ON U.S. FIRMS DOING BUSINESS IN VIETNAM.

(On February 3, 1994 President Clinton announced that the embargo on trade with Vietnam would be lifted.)

2. THE PRESIDENT SHOULD INITIATE A PROCESS LEADING TO NORMAL DIPLOMATIC RELATIONS WITH THE PEOPLES REPUBLIC OF VIETNAM.

Diplomatic recognition would provide the necessary presence in Vietnam to facilitate engagement across a range of sectors, politically, economically, and culturally, that would ensure the benefits that can flow from removal of the embargo. It makes little sense to lift the embargo unless we take the necessary follow up steps so U.S. businesses, including small and medium sized businesses, can compete effectively in Vietnam. Withholding the kind of services that would be available following full diplomatic recognition would significantly limit the benefits of engagement with Vietnam.

3. U.S. TRADE PROGRAMS ADMINISTERED BY AGENCIES SUCH AS OPIC, THE EXPORT-IMPORT BANK, U.S.AID, THE TRADE DEVELOPMENT ADMINISTRATION, U.S./ASIA ENVIRONMENTAL PARTNERSHIP AND THE U.S. TRADE NEGOTIATOR SHOULD MOVE AT ONCE TO INCLUDE VIETNAM AS AN ELIGIBLE PARTICIPANT. U.S. COMPANIES THROUGH TRADE MISSIONS, TECHNICAL ASSISTANCE AND APPROPRIATE FINANCIAL INCENTIVES, SHOULD BE ENCOURAGED TO INVEST IN VIETNAM.

Economic development in Vietnam is about to take off, fueled by investments by international financial institutions and by privately funded efforts of every significant nation but the U.S. The U.S. needs to review the adequacy of its programs for assisting American business, particularly small and medium sized business, to participate in this development.

4. PROGRAMS SUPPORTING FOREIGN STUDENTS STUDYING IN U.S. COLLEGES AND UNIVERSITIES SHOULD BE OPEN TO VIETNAMESE STUDENTS ON A NONDISCRIMINATORY BASIS.

The experiences of students in the United States from other Asian nations have contributed to the advancement of reform efforts in these nations. There is every reason to believe that the same result will be achieved with the Vietnamese. U.S. technical and managerial skills are highly valued in Vietnam. The majority of the delegation believes that engagement in educational exchange and in efforts to develop in the Vietnamese the skills needed for a market economy would be an important investment and would produce benefits for both countries.

Currently only about 60 Vietnamese nationals—including 15 Fulbright scholars—study in the United States. The terms of the U.S. embargo do not allow Vietnamese to study technical subjects. The majority of the delegation strongly recommends that the President review the rationale for restricting the number and curricula of Vietnamese studying in the U.S. with the aim of increasing educational opportunities for Vietnamese students in U.S. colleges and universities as a mechanism to assist in the improvement of conditions in Vietnam.

5. U.S. DEPARTMENT OF ENERGY LABORATORIES SHOULD BE ENCOURAGED TO PROVIDE TECHNOLOGY TRANSFER AND TECHNICAL AND MANAGEMENT ASSISTANCE IN DEVELOPING THE SCIENCE, TECHNOLOGY AND ENERGY INFRASTRUCTURE OF VIETNAM.

Present electric power capacity for 74 million Vietnamese is only 4 Gwt, or about 0.05 kw per capita. U.S. electric generating capacity is more than 60 times larger—over 3.0 kw per capita. In Ho Chi Minh City the delegation was told of a recent 4-month period during which power blackouts of the city for 5 to 6 days a week virtually stopped the economy.

Sixty percent of Vietnam's electric capacity is hydroelectric. Because Vietnam hopes to have 6 Gwt on line by the year 2000, a significant expansion of 300 to 350 megawatts (Mwt) per year ($500-$600 million per year) will be required over the next decade. The Vietnamese are planning to expand coal capacity in addition to expansions of hydroelectric power to meet this demand. The delegation was asked for help securing U.S. assistance in underground and open pit coal mining technology, coal preparation, transportation and port facility development (Vietnam exported 1 million tons of coal last year). The delegation was also asked for help in obtaining used electric generation equipment and facilities for testing electric generation equipment.

6. U.S. OIL AND GAS PRODUCTION, PROCESSING AND SERVICE COMPANIES SHOULD BE ENCOURAGED TO ASSIST VIETNAM IN DEVELOPING ITS OIL AND GAS RESOURCES AS AN IMPORTANT SOURCE OF ENERGY AND FOREIGN EXCHANGE.

Vietnam is cooperating with many foreign oil companies in developing its offshore reserves, but while the embargo is in place, no American firms may participate in development and production. The delegation was told that the Vietnamese very

82-926 O - 94 - 4

much want to use American technology and expertise to develop their oil reserves as a basis for the development of the economy. Vietnam produces 130,000 barrels per day currently. Oil generates significant foreign exchange earnings representing a substantial percentage of the government's revenues. Based on the reserves known today, production of oil by Vietnam is projected to double sometime in the first decade of the 21st century and continue to double again before 2020 to the great benefit of Vietnam's plans for economic development. Refined product demand in Vietnam is projected to remain below crude oil production for the foreseeable future. Vietnam is presently considering whether to continue to import refined petroleum products, to construct domestic refineries for its own oil or to refine imported oil for domestic consumption. In any of these instances, American technology and expertise would be very competitive. In addition, there are good prospects for development of natural gas for domestic consumption in the short run and for possible liquid natural gas exports should that market develop in Japan, China and elsewhere in the western Pacific Rim.

Oil and gas development is an area where the advantage of American technology is well known and eagerly sought by the Vietnamese. The U.S. should press this advantage.

7. RESTRICTIONS ON U.S.AID PROGRAMS IN VIETNAM SHOULD BE RELAXED TO PERMIT A BROADER DEVELOPMENT EFFORT TO IMPROVE THE LIVES OF THE VIETNAMESE PEOPLE.

The delegation's visit to the Central Vocational School for Handicapped Persons No. II in Thu Duc in the suburbs of Ho Chi Minh City reminded everyone that the need for humanitarian assistance in Vietnam is far from satisfied by current aid programs. This school was converted from a vocational school serving the general population to a vocational school for the handicapped in 1975. The school has taught repair of electric and electronic appliances, motorcycle repair, embroidery, dressmaking and carpentry. Since May 1993 with the assistance of U.S.AID and American medical personnel the school has manufactured artificial legs. Many of the recipients have war-related injuries either dating from the war or from injuries caused by mines and bombs after the war's end. This program could be significantly expanded and, in the opinion of the majority of the delegation, should receive continued U.S.AID assistance. More generally, the President should consider substantially increasing the role of the United States in humanitarian aid to Vietnam beyond the half dozen projects currently supported in part by U.S.AID.

8. ANY VIETNAMESE LIVING IN VIETNAM WHO RECEIVED RECOGNITION FROM THE U.S. GOVERNMENT FOR VALOR DURING THE WAR IN VIETNAM SHOULD BE ELIGIBLE TO ENTER THE UNITED STATES.

The delegation was deeply impressed by the experiences of Major An Quy Nguyen, a decorated war hero who flew helicopters in combat for the South Vietnamese (Republic of Vietnam) Air Force. Major An was awarded the Distinguished Flying Cross by the President of the United States for valor under fire in disregard for his own safety resulting in the rescue of four American soldiers whose helicopter had been shot down. In a later engagement, Major An stayed at the controls of his own helicopter after it had been hit and caught fire and landed the ship and its crew safely. However the burns suffered in this action resulted in the amputation of both Major An's arms.

After the Republic of Vietnam was defeated in 1975, Major An was arrested and sent to a reeducation camp. He was released soon thereafter because officials at the camp judged him "useless" due to the loss of his arms. In subsequent years Major An suffered greatly in the Peoples Republic of Vietnam. His family was separated and some members lost. His repeated attempts to leave the country attracted unfavorable attention from the Vietnamese government. He also found no relief in the procedures set up by the U.S. government to address the situation of Vietnamese who fought with the U.S. during the war. Major An does not qualify for the U.S. Orderly Departure Program reeducation camp subprogram because his time in the camp was too short to meet the criteria of the regulations. Apparently his situation does not fit very easily into any other category allowing emigration to the United States.

Happily, the efforts of a few determined individuals proved more effective than the collection of federal programs set up to address the condition of persons like Major An. Over several years Colonel Noboru Matsuoka (Ret) of Moffett Field, California, searched for and located Americans who had served with Major An, including some whose lives he saved. One of these, Col. John Liner, lives in Leesville, Louisiana. Led by Col. Matsuoka, a campaign was mounted to bring Major An to the United States. Many telephone calls, letters and visits to U.S. authorities over several years culminated in granting humanitarian parole to Major An and his daughter by the U.S. Immigration and Naturalization Service. This action offered an opening for Major An and his daughter to leave Vietnam for the U.S. if the Vietnamese would agree to it and the means could be provided. Before the delegation's departure, members had been made aware of Major An's situation. The delegation raised Major An's case with the Vietnamese government at the highest levels. As a result of this intervention, Major An and his daughter were allowed to leave Vietnam on the delegation airplane. Major An and his daughter currently are being sponsored in California by Col. Matsuoka, where they may stay for a year before applying for adjustment in immigration status.

The delegation believes that Vietnamese decorated for valor by the U.S. government during the Vietnam War should not have to go to the lengths required in Major An's case or to wait as long he did before being admitted into the United States. Provisions should be made for much more humane treatment of those Vietnamese who may be suffering because of their service alongside the U.S. during the war.

III. THAILAND

Visits

In Thailand, the delegation visited Bangkok and the ancient capitol of Ayutthaya. Upon arrival, the delegation was hosted at dinner by Deputy Prime Minister Dr. Amnuay Viravan, who has responsibility for energy and environmental issues in the government of Prime Minister Chuan Likphai. The next day, Ambassador David E. Lambertson and his country team provided the delegation with a full briefing on U.S.-Thai relations. The delegation then met with the Prime Minister and his cabinet and later that afternoon had an audience with His Royal Majesty Bhumibol Adunyadet, the King of Thailand. That evening the delegation drove to the 15th century capital at Ayutthaya outside Bangkok for a dinner hosted by Her Royal Highness Queen Sirikit Kitiyakara and attended by the Thai Royal family.

Findings

1. THE U.S.-THAI RELATIONSHIP IS WARM, STRONG AND CHARACTERIZED BY MUTUAL RESPECT

The United States and Thailand have a long history of friendship and cooperation. The first U.S. trading ship arrived in Thailand (then Siam) in 1821. The United States entered into a Treaty of Amity and Commerce with Siam in 1833, the first treaty between the U.S. and an Asian nation. Relations have been respectful and friendly for nearly all of the 160 years since then.

After the Japanese occupied Thailand in 1941 the Bangkok government declared war against Great Britain and the United States. However, the Thai Minister to the United States refused to deliver the declaration of war to the U.S. government because he did not believe it reflected the will of the Thai people. After World War II the U.S. helped to shield Thailand from demands of other allies for reparations. Thailand sent troops to Korea and supported U.S. policies in Vietnam and during the Persian Gulf War.

Recently, relations between the U.S. and Thailand worsened dramatically and have only recovered in the last year. In February 1991 the Thai military took power from a democratically elected government. The U.S. denounced the coup and, as required by law, cut off assistance programs and military cooperation. Violent suppression of non-violent pro-democracy protests in May 1992 by untrained troops with loaded M-16 rifles further distanced the Thai government from both the U.S. and international public opinion.

This tragedy horrified the great majority of Thais and galvanized the King into a rare intervention into Thai politics. Under the King's guidance, the nation returned to democracy and civilian rule in September 1992 with the free election of the parliament from which Prime Minister Chuan formed his current five-party coalition government.

Chuan is attempting with some success to return Thailand to the politics of normalcy. The nation is prosperous. Economic growth has been 8 percent per year for the last three years. The Thai GNP doubled between 1987 and 1992. Chuan appears to be succeeding in his goal of demonstrating that democratic government can be honest and accountable.

The principal political goals of the Chuan government are to develop respect for democracy at the grass roots and consensus processes at the national level and help the Thai military emphasize professionalism and find a relationship to civilians government appropriate to a modern economic state. Chuan needs to demonstrate respect for the rule of law. A longer term goal is to pursue a constructive engagement with Myanmar (formerly Burma), with whom the Thais share a very long border.

Social problems on the Chuan agenda include:

- enforcement of child labor and prostitution laws;
- increasing opportunities for women;
- developing a national response to a serious AIDS epidemic;
- reducing the drug traffic from Laos and Burma through Thailand;
- narrowing income disparities between rural and urban populations; and
- increasing opportunities for labor to organize.

There is a sense that the Chuan government is making serious attempts at addressing these issues, is gaining respect from a substantial portion of the Thai political spectrum and deserves U.S. support.

2. NOW IS A GOOD TIME TO ADDRESS THE TRADE PROBLEMS BETWEEN THE U.S. AND THAILAND

The climate of warmth between the U.S. and Thailand offers promise that U.S. concerns with trade can be successfully addressed. These concerns are:

Intellectual Property Rights—Thailand needs to enact and enforce strong laws bringing protection of intellectual property in Thailand up to international standards. Such action will benefit both foreign companies and legitimate Thai businessmen who hold licenses and are harmed when illegal copies flood the market. U.S. and other exporters to Thailand need to be aware that setting prices for imports in Thailand far above costs makes piracy attractive.

Trade Deficit—Two-way trade between the U.S. and Thailand was approximately $12 billion in 1993 with Thai exports leading and a U.S. trade deficit of over $3.5 billion. The U.S. is the largest Thai export market. U.S. exporters of sophisticated goods such as energy and telecommunications systems, environmental technologies and computers and related equipment are very competitive in Thailand and could contribute to reducing the trade deficit the U.S. currently runs with Thailand. Thailand's electric capacity is 10 Gwt; the planned capacity for 2000 is 19 Gwt. Currently Japan is the leading investor in Thailand. The U.S. is second. Japan, which supplies substantial financing to its export companies, has four times as much investment in Thailand in electric power equipment as the U.S. because of limitations on U.S. Export Import Bank funds. By operation of law, U.S. Trade and Development Administration (TDA) assistance, which funds feasibility studies, was suspended after the military coup in 1991 and was only reinstated in 1993. TDA has been one of the more successful tools the U.S. has used in opening new markets for American companies by enabling these companies to become involved in the planning stages of infrastructure projects. As with the Export Import Bank, funding levels for TDA are limited. In fiscal year 1993 only $9.6 million was available for all of Asia. This amount could support only 50 projects for the entire region. The majority of the delegation believes that substantially reducing the U.S. trade deficit with Thailand (and other rapidly growing Asian economies) may require fundamental restructuring of these and other U.S. export promotion programs.

Drugs—Most of the world's heroin passes through Thailand from Laos and Myanmar. (Small but significant amounts also are transported through China and Vietnam.) Recent seizures in Thailand of roughly 1,000 kilograms (worth $1 billion) indicate that trafficking may be increasing. Two private armies of 15,000 to 20,000 men in northeastern Myanmar produce 70 percent of the heroin sent to the U.S. The Thais have been very cooperative with the U.S. DEA in attempting to combat the flow of drugs through Thailand. The value of this cooperation only underlines the need for the U.S. to establish some sort of relationship with the government of Myanmar to develop an effective policy of interdiction at the source.

Worker Rights—Draft legislation introduced in the recent session of Parliament is acceptable to Thai labor leaders and should be enacted later this year.

Tax Treaty—The conclusion of a tax treaty between the U.S. and Thailand is the highest priority of U.S. business in Thailand. Thailand already has tax treaties with 26 other nations. The lack of a treaty puts U.S. companies at a competitive disadvantage.

Recommendations

1. THE U.S. SHOULD CONTINUE TO SUPPORT AND ENCOURAGE THE THAI GOVERNMENT IN ITS EFFORTS TO DEVELOP THE THAI ECONOMY1 ESPECIALLY THE CONCLUSION OF A DOUBLE TAXATION TREATY TO REMOVE DISINCENTIVES FOR U.S. BUSINESS OPERATING IN THAILAND.

2. THE U.S. SHOULD REVIEW ITS EXPORT PROMOTION PROGRAMS TO DETERMINE WHAT CHANGES CAN BE MADE TO ENCOURAGE U.S. BUSI-

NESS, ESPECIALLY IN THE ENERGY AND ENVIRONMENT SECTORS, TO INVEST IN THAILAND.

3. THE U.S. SHOULD COOPERATE AS FULLY AS POSSIBLE WITH THE THAI GOVERNMENT IN INNOVATIVE PROGRAMS TO COMBAT THE SPREAD OF AIDS IN THAILAND AND ATTEMPT TO LEARN AS MUCH AS POSSIBLE FROM THE DISTURBING EPIDEMIC THE THAIS ARE CURRENTLY EXPERIENCING.

4. THE U.S. SHOULD TERMINATE ACTIVE REVIEW OF THE GENERALIZED SYSTEM OF PREFERENCES (GSP) TRADE PROGRAM WHEN THE THAI PARLIAMENT ENACTS (AS EXPECTED LATER THIS YEAR) APPROPRIATE LEGISLATION TO PROTECT THE RIGHTS OF WORKERS IN STATE-OWNED ENTERPRISES.

5. THE ADMINISTRATION SHOULD CONSIDER MAKING OFFICIAL CONTACT, SHORT OF RECOGNITION IF NECESSARY, WITH THE GOVERNMENT OF MYANMAR TO EXPLORE DEVELOPMENT OF COORDINATED POLICIES TO REDUCE THE DRUG TRAFFIC THROUGH THAILAND.

IV. INDONESIA

Visits

In Indonesia, the delegation visited the cities of Jakarta and Yogyakarta on the island of Java and the provinces of Bali and Irian Jaya. Upon arrival in Jakarta, the delegation received a briefing from Ambassador Robert Barry and his country team. Subsequently, the delegation met with Minister of Foreign Affairs Ali Abdullah Alatas, Minister of Trade Satrio Budihardio Joedono and Parliamentary Deputy Speaker John Ario Katili. Later, the delegation met and had dinner with Mines and Energy Minister Lt. Gen. Ida Bagus Sudjana and representatives of American companies doing business in Indonesia. Representatives of the delegation visited two projects funded in part by U.S.AID: Project Rescue, which works with Indonesian street children, and the Yayasan Kusuma Bhuana Family Planning Clinic located in a suburb of Jakarta, which provides health services to women and children on a sliding fee basis.

The following morning, the delegation met with President Soeharto and then departed for Yogyakarta. In Yogyakarta, one of three designated special provinces, the delegation met with Sultan Hamengkubuwono X of Yogyakarta, and had lunch with the Sultan and Queen Kanjeng Ratu Hemas.

In Bali, the delegation visited the Bamboo Training Center in Penglipuran, run by the Indonesian Environmental Bamboo Foundation, a not-for-profit organization which has received support from U.S.AID. With this support, the Foundation recently discovered a process which will prolong the natural life of bamboo and is currently undertaking research on various uses of bamboo as well as marketing techniques of this renewable resource which many believe may prove an alternative to wood derived from Indonesia's and the world's fragile rain forests. In addition, members of the delegation visited Dana's Company, an Indonesian prefabricated house construction company typical of those small businesses whose exports benefit from the Generalized System of Preferences (GSP) which allows duty free entry of certain imports into the United States. This favorable tariff treatment is currently under review by the U.S. Trade Representative. Prior to leaving, the delegation met with the Deputy Governor of Bali.

The delegation also visited the province of Irian Jaya, Indonesian New Guinea, the location of the P.T. Freeport Indonesia's joint U.S.-Indonesian development at Tembagapura. The delegation visited the Grasberg copper and gold mine, the world's single largest gold reserve and one of the world's largest open pit copper reserves. Over $230 million in goods, equipment and services are procured annually in the United States to support this mining operation.

Findings

1. THE UNITED STATES REMAINS LARGELY UNAWARE OF INDONESIA'S SIZE AND DIVERSITY

Although Indonesia is the fourth largest nation in the world, with the world's largest Muslim population, far too little is known in the United States about this vast, resource-rich country located at the strategic crossroads connecting the Pacific and Indian Oceans. Stretching across 3,000 miles, Indonesia is composed of over 17,000 islands (13,000 of which are inhabited), with the preponderance of the population concentrated on four provinces: Sumatra, Java/Madura, Kalimantan and Sulawesi. Over 100 million people, or 60 percent of Indonesia's total population, reside on Java/Madura, which has less than 7 percent of Indonesia's total land mass. Java has the highest population density rate in Indonesia: 814 persons per square kilometer, one of the world's most dense. In contrast, less than two million people reside on Irian Jaya, less than one percent of the total population. Irian Jaya contains almost 20 percent of the archipelago's land mass, but has a population density rate of only 4 persons per square kilometer. More than 300 different ethnic groups speaking about 583 languages and dialects have been identified in Indonesia, although Bahasa Indonesian, which was introduced as the lingua franca following independence, is widely spoken throughout the country.

The delegation's visits to Java, Bali and Irian Jaya gave us some appreciation of these differences, and the difficulties they pose for the delivery of services by the central government. One of the more striking contrasts was presented by the delegation's visit to Jakarta, a major metropolitan area with over 8 million people, and the visit to Irian Jaya, two time zones and a six-hour flight away, which is sparsely populated with little infrastructure development other than what has been brought by commercial operations.

2. INDONESIA'S ECONOMIC TRANSFORMATION HAS BEEN VERY IMPRESSIVE AND HAS REACHED ALL SECTORS OF THE POPULATION

Since 1967, Indonesia has experienced an economic miracle: in 1970, over 60 percent of the population (roughly 70 million people) lived at or below the poverty rate; today that rate has been reduced to 15 percent (approximately 27 million people). In 1965, Indonesia's per capita income was approximately $50 per year, about half that of Bangladesh, Nepal or Nigeria. In 1989, per capita income exceeded $500 annually and by the year 2000, it is expected that the average per capita income will reach $1000 per annum. In 1990, the World Bank found that from 1970 to 1990 Indonesia achieved the highest annual average reduction in the rate of poverty of all countries studied. The World Bank has also found that Indonesia's is one of the ten fastest growing economies in the world, and has excellent prospects for sustaining growth levels at or above 6 percent annually through the end of the decade.

The benefits of this notable achievement have been felt across all sectors. A study completed last year by the World Bank, entitled *Indonesia: Sustaining Development,* noted that there has been a gradual reduction in disparities—both between rich and poor and among Indonesia's regions. Today Indonesia's distribution of wealth is roughly equivalent to that in the United States, and both access to and utilization of health and other social services by the poor have increased dramatically.

By every measure, substantial progress in human resource development has been achieved. Life expectancy has increased markedly: for men, from an average of age 43 in 1965 to age 60 in 1989, and for women, from an average of 45 in 1965 to age 63 in 1989. Infant mortality rates declined significantly: from 128 deaths per 1,000 lives births in 1965 to 64 per 1,000 in 1989. The maternal mortality rate dropped to about 800 per 100,000 live births by 1980.

A small but important part has been played by U.S. assistance in improving the health of women and children, through training programs in family planning and health including a five year $20 million grant now being implemented to train nurses, midwives and physicians. Substantial progress has been made in reducing the shortage of physicians and nurses, as some members of the delegation saw during a visit to the Yayasan Kusuma Bhuana Family Planning Clinic established in a suburb of Jakarta in part with U.S. assistance. Nonetheless, in 1984 there were 1,260 people per nurse in Indonesia, compared to the U.S. average of 310 to 1 and a worldwide average of approximately 900 to 1. While major strides have been made in decreasing the population per physician ratio (over 46,000 to 1 in 1960), more remains to be done to reduce the 1984 rate of approximately 9,400 to 1, substantially above the average of the East Asian and Pacific and almost four times that of India (roughly 2500 to 1). The delegation noted that joint ventures involving U.S. companies have helped make medical personnel available to indigenous populations in remote areas such as Irian Jaya, where P.T. Indonesia Freeport operates free clinics for residents in the area surrounding the Freeport development. Typically, the Freeport clinics serve more than 5,000 patients each month. This is one of the tangible benefits which has resulted from the presence of international corporations in these areas.

Dramatic strides have also been made in increasing the daily caloric supply (which increased from 1,796 calories per capita in 1965 to 2,670 calories per capita in 1988). The delegation noted that, on Irian Jaya, efforts are underway to help diversify diets through the introduction of new crops. These efforts were sponsored in part by the Government of Indonesia and in part by the private sector to help assure that more nutritious calories are consumed.

Universal primary education has been achieved and education is now mandatory for both girls and boys through age 12. Just over 46 percent of students enroll in middle school. Since 1950, the literacy rate amongst adults has increased dramatically from below 20 percent to about 74 percent today. Indonesia has been one of the leaders in the developing world in increasing educational opportunities for women. Female school enrollments have risen faster than average, with the result that 48 percent of all primary school students and 45 percent of secondary school students are female. In addition, substantial increases have been made in the enrollment rate for the neediest 40 percent of children, more than doubling at the secondary level for females and increasing by more than half for males.

3. FUTURE DEVELOPMENT CHALLENGES IN INDONESIA ARE SIGNIFICANT

Many challenges remain and in part will be more difficult to meet, because past successes have increased expectations of quick and sustained results.

Despite the dramatic reduction in the poverty rate, 27 million people, or 15 percent of the population, remain in poverty. Many more live near the poverty level. Achieving further reductions will be difficult, as many in these groups live in remote, isolated areas with little infrastructure and where human resource development has not reached the national level. According to a recent World Bank study the incidence of poverty ranged from 45.6 percent in East Nusa Tenggara to 1.3 percent in DKI Jakarta. In addition, hard core pockets of poverty persist in some of the major metropolitan areas of Java, including Jakarta, and these will be difficult to reduce in the short run. Targeted intervention programs such as Project Rescue, a program for street children in Jakarta supported in part with funds from U.S. AID, and social and business development programs such as those sponsored by P.T. Indonesia Freeport in Irian Jaya will be necessary if significant strides are to be made in the next decade in further reducing the poverty rate. Even so, results will not be produced quickly since skills development will be as important as the delivery of services in this effort.

Much of Indonesia's past success is attributable to the extraordinarily high growth level which was sustained over the 1970 to 1990 period, averaging some 7 percent annually. For further success the current 6 percent growth rate must be sustained, and that will also be difficult although certainly achievable. Further development must concentrate on export markets and continuing the diversification of the economy that has been underway since 1986.

Growth and the transformation of Indonesia from a primarily agrarian economy to an industrial economy have produced environmental challenges to the quality of life in Indonesia. Many in Indonesia do not have access to safe, clean drinking water; unsafe water is a major cause of widespread public diseases including diarrhea, typhoid and cholera. In Jakarta alone, the costs of mortality associated with diarrhea are estimated at roughly $200 million annually. A 1989 survey found over 93 percent of Jakarta's shallow wells to be contaminated, a serious situation, since only 15 percent of all urban household are connected to public water systems and therefore rely on wells for water. In addition, few cities throughout Indonesia have public sewage systems, or even the beginning of such systems. In Jakarta some 30 percent of all solid waste is uncollected.

Although air pollution does not yet approach levels of Bangkok or Beijing, the lack of an adequate public transportation system in Jakarta and over reliance on automobiles is becoming more serious. Health-related costs from the use of leaded gasoline are estimated at $50 million annually in Jakarta, while costs associated with pollution from suspended particulates (emitted by power plants and other industrial facilities) is approaching $75 million yearly. Some studies have projected that, absent corrective action, air pollution in Jakarta and other urban areas could increase by 250 percent in the next seven years.

Industrial pollution is also increasing. The World Bank estimates that, absent corrective actions, biological oxygen demand (BOD) levels from such pollution will increase in Indonesia by 250 percent by the year 2000 and toxic and heavy metal wastes will increase 280 percent over that same period. Major investments in pollution abatement will be needed to prevent serious health problems.

The Government of Indonesia has accorded a high priority to environmental protection and has taken a number of steps to improve environmental management and protection. An Environmental Protection Agency (BAPEDAL) was established, and environmental impact analysis requirements for development projects have been promulgated. These initiatives and others have laid a good foundation for future initiatives to abate and respond to environmental challenges in the years ahead.

Indonesia has a predominantly young population: over 36 percent of its people were under the age of 14 in 1989. One of the most efficient voluntary birth control programs in the world (which was developed and implemented with substantial U.S. government support) has reduced the population growth rate from its average of 2.4

percent from 1965 to 1980 to 1.6 percent today. Nevertheless, Indonesia faces a daunting task of providing jobs for the over 2.3 million new entrants coming into the labor market each year.

The Indonesian Government has made a commendable investment in education and particularly in expanding higher education facilities. Ten percent of all central government expenditures were devoted to education in 1989, up from 7.4 percent in 1972. Throughout Indonesia there remains a continuing need for greater post-primary education skills, particularly in technical and management areas. U.S. companies have played a small but important role in providing education and training opportunities, particularly in rural areas. The delegation observed an excellent example of this at the P.T. Freeport Indonesia development on Irian Jaya.

4. INDONESIA PROVIDES SUBSTANTIAL OPPORTUNITIES FOR U.S. EXPORTS

With a GDP of some $139 billion, and an economy expanding at more than 6 percent each year, a rate projected to be sustained for the remainder of this decade, Indonesia's potential market presents significant opportunities for American companies. Over the next decade Indonesia is expected to import infrastructure-related projects valued at $113 billion. By merely maintaining our current market share, U.S. exports could double to almost $6 billion annually. But because the major growth sectors (aerospace, telecommunications, environmental technology, energy, transportation and services) are in areas in which the United States is highly competitive, this rate can be exceeded if the Government of Indonesia continues its encouragement of foreign investment. Trade liberalization, including the reduction of nontariff barriers, import monopolies and licensing restrictions, has been critical to attracting such investment. The delegation raised a number of concerns of U.S. companies and U.S. officials about the need for improved intellectual property protection, continued deregulation, further privatization, and improved incentives in the huge oil and gas sector. These concerns were taken under advisement, and if acted on could improve the attractiveness of the Indonesian economy to U.S. investors and businesses.

The delegation was told by the U.S. Department of Commerce that particularly good prospective markets for U.S. trade and investment are iron and steel products; industrial chemicals; oil and gas field machinery and services (40 U.S. companies currently operate in Indonesia under production sharing contracts; since the world oil price decline of 1984-86, production has steadily increased and is projected to grow at 16.5 percent per annum over the 1993-95 projection period); electric power systems (with an estimated annual growth rate of an astonishing 17.5 percent through 1995, U.S. exports in this sector are expected to grow by 8 percent per year; 80 power projects at a total cost of $12 billion are expected to be constructed by the end of this century); pumps, valves and compressors; textile machinery and equipment; plastic materials and resins; construction equipment; machine tools and metal working equipment; building products; commercial vessels and equipment; nonferrous metal; pulp and paper machinery; forestry and woodworking equipment; laboratory and scientific instruments; materials handling machinery; aircraft and parts; computer software; pollution control equipment; printing and graphic arts equipment; medical equipment; and business equipment (non computer).

Despite this growing market, the share of U.S. participation is still relatively small compared to that of other nations and has declined in recent years. A concerted effort is now being made by U.S. trade officials and the private sector to reverse this trend. It is apparent however that additional trade missions from the United States are needed, particularly if small and medium sized businesses are to increase their share in the many potential opportunities which Indonesia's future infrastructure plans to provide. Moreover, the United States should review its current credit programs to assure that U.S. companies seeking to do business in Indonesia are not at a competitive disadvantage to companies from other nations.

The United States and Indonesia must cooperate to resolve market access questions and build on the momentum established by the Seattle APEC summit and the approval of a new revised General Agreement on Trade and Tariffs (GATT). The United States has much to offer Indonesia in technology and state of the art equipment and training. Indonesia offers significant opportunities for U.S. businesses and the creation of jobs both in the U.S. and in Indonesia.

The economies of Indonesia and the United States are complementary. The delegation was encouraged by the genuine desire of Indonesian officials to see linkages between our economies increased and believes the American market share in Indonesia could increase substantially in the future.

5. THE U.S. HAS A DEEP INTEREST IN A BILATERAL SECURITY RELATIONSHIP WITH A STABLE AND FRIENDLY INDONESIA

As important as improving our economic relationship with Asia is creating the conditions for continued peace and mutual security in Asia and the Pacific. The importance of our security relationship has increased with the end of the Cold War.

Indonesia's strategic location, sprawling across over 3,000 miles of vital sea lanes in Southeast Asia which connect the Pacific and Indian Oceans, accords a high priority to maintaining a friendly and cooperative relationship. The United States has a strong national interest in exercising internationally recognized transit and overflight rights through and above these sea lanes.

In addition, its location has placed Indonesia at the forefront of a number of potentially destabilizing regional disputes, such as the competing claims of China and Vietnam to the Spratly Islands and the recent turmoil in Cambodia. In both these controversies, Indonesia has taken the lead in seeking a peaceful, mediated solution, an approach the United States has strongly supported and which is in our interest. Indonesia has been active in promoting the new ASEAN Regional Forum, in which the United States also participates and which will be important to the future stability of the region.

In other security and foreign policy areas, including nonproliferation and international arms control, the United States and Indonesia have worked cooperatively. Moreover, as leader of the NonAligned Movement and with the world's largest Muslim population, Indonesia's is an important and weighty voice in Islamic organizations. In these and other fora, since 1967 Indonesia has consistently been a voice of moderation, pragmatism and cooperation, which has been important to the United States and to stability and peace in Asia. Nothing could be more antithetical to the interest of the United States than a return to instability in Indonesia. Renewed turbulence combined with the increasing interdependence of economies in the region and the world would inevitably have serious, adverse impacts beyond Indonesia's borders.

As Admiral Larson, Commander in Chief of the Pacific forces, pointed out to the delegation in a briefing paper, the United States has a strong interest in maintaining good military-to-military contacts with Indonesia, which is a component of our forward based deployment in the Asia-Pacific theater. The Indonesian military, known as ABRI, plays an important role in the social, political and economic development of the country. Active duty ABRI officers serve in a variety of positions: in State-owned corporations, in philanthropic foundations, in Parliament, and at all levels of civic government.

The delegation takes note of the strong objections that U.S. diplomatic and military officials have registered to attempts in the Congress to limit the access of Indonesia's military to U.S. weapons, defense systems and military education and training programs. The cut off of grant assistance to support such activities in 1992 has not changed the military for the better. To the contrary, it is the view of U.S. officials that decreased access will work against the U.S. objective of democratizing the military and sensitizing the military to the importance of the observance of human rights and the rule of law as well as adherence to democratic principles. The Embassy pointed out that Indonesia's top three defense officials—the Defense Minister1 the Commander in Chief of ABRI and his Chief of Staff—are IMET graduates and are dedicated to a professional Indonesian military and to closer and productive defense relations with the U.S.

The delegation notes that no official in the Indonesian government requested that grant assistance for the purposes of military education and training be resumed. However, further restrictions on commercial sales of military equipment were raised and are perceived as unfair and unjustified. In the end, these restrictions will not keep Indonesia from acquiring new equipment. Instead, Indonesia will seek other markets and sources, weakening the link with the United States and substantially reducing our ability to influence and encourage change.

6. EMERGING POLITICAL CHANGE IN INDONESIA WILL REQUIRE INTELLIGENT RESPONSES BY THE U.S.

The delegation notes that domestic political change has not kept pace with Indonesia's economic reforms or emerging innovative regional economic and security initiatives. The political system remains highly centralized and tightly controlled. In large part this system is a result of the high premium the leadership and most Indonesians have placed on the need for stability in this land of 300 ethnic groups dispersed over 13,000 islands.

At the same time, the delegation takes note of a number of positive changes. Newspapers in Jakarta, or at least the English language press, were full of articles expressing critical views about various government programs and proposals. It was apparent that there is an emerging independent press with open debate over ideas and other concerns. Criticisms were freely offered by students and other activist groups and associations. Both the number and range of activities undertaken by nongovernmental organizations (NGOs) have increased, particularly on environmental issues. The delegation was told that fewer Indonesians have been prohibited from traveling abroad, and in January regulations were promulgated to limit the role of the military in strikes and to allow workers to organize with negotiating authority at the local plant level.

From the standpoint of the United States, Indonesia can and should do more. It is important to recognize, however, the great strides Indonesia has made in improving the quality of life for all of her citizens, which will lay the groundwork for further changes in the future based on a stable and solid middle class.

Recommendations

1. RENEW GENERALIZED SYSTEM OF PREFERENCES (GSP) BENEFITS FOR INDONESIA.

Favorable tariff treatment by the United States for certain Indonesian goods will not impact Indonesia's economy in a significant way. Only about 13 percent of these goods are exported to the United States, and of these exports only about 14 percent receive duty free treatment.

However, certain sectors of the Indonesian economy will be hit hard. Ironically, furniture and handicrafts, sectors dominated by small businesses and cooperatives, could be the hardest hit and the least able to compete absent GSP treatment, since their competitors for the most part receive GSP. One company whose future is at risk in Klaten began with support from the National Cooperative Business Association, a U.S. association, and has grown from a total work force of 35 in 1985 to 1,500 today. While earnings were approximately $1.4 million in 1991, the profit margin is sufficiently small that the loss of the tariff preference could mean the difference in this company's survival, as well as the survival of the 1,500 jobs it supports. Many of these workers have come from the rural agricultural sector and have been pulled out of poverty through these jobs. It would be a travesty if GSP were revoked for the reason that insufficient progress had been made in improving workers rights in general, and that revocation caused workers in small businesses to be laid off.

The delegation notes that a number of U.S. companies have also expressed concern about the impact that suspension, or even a continuation of the GSP review for another year, could have on commercial opportunities in Indonesia. Suspension of GSP would result in the suspension of Overseas Private Investment Corporation (OPIC) programs, and potentially could impact some $800 million in pending applications and another $450 to 500 million in the FY 1994 pipeline. (New projects only would be suspended.) In addition, it is clear that suspension or continuation of the review would be a major irritant in our bilateral relationship and could lead to lost contracts worth billions of dollars in the energy, telecommunications, transportation and other sectors.

2. THE U.S. SHOULD SEEK IMPROVED MILITARY TO MILITARY CONTACTS WITH THE INDONESIAN ARMED FORCES.

The majority of the delegation believes it is important to improve military to military contacts if the United States is to have the ability to influence change in a positive way. Attempts to restrict such contacts further should be opposed.

In addition, the majority of the delegation believes that the Administration should initiate high level contacts with the Indonesian military to show that it believes the relationship is important. In light of the uncertain situation in the North Pacific, and the potential for destabilizing regional disputes in the area, a close and cooperative relationship is in our interest and should be nurtured, not ignored.

3. THE ADMINISTRATION SHOULD REVIEW U.S. PREFERENTIAL CREDIT PROGRAMS TO ENCOURAGE U.S. BUSINESS TO COMPETE IN INDONESIA.

The delegation heard from both U.S. companies and from Indonesian officials that the current U.S. credit system is not competitive with that offered by other nations and places U.S. companies at a disadvantage, even though their products are of high quality and reliability.

Therefore, the majority of the delegation urges the Administration to undertake a review of how our current export programs have been implemented and whether they are sufficient to enable U.S. companies to compete effectively.

4. THE U.S. SHOULD SUPPORT PROJECTS AIMED AT TARGETED ASSISTANCE TO THE POOR AND ECONOMIC DIVERSIFICATION.

The delegation was impressed by the work U.S.AID is undertaking to support assistance for the needy in Jakarta, notably Project Rescue, and urges support for similar projects in the eastern provinces and other underdeveloped parts of Indonesia. Further reductions in the poverty rate will be difficult in these areas, and we should support efforts to identify and reach out to the underserved in areas such as Timor, and East Nusa Tenggara.

The delegation also notes that, with 2.3 million new entrants coming into the labor force each year, it is critically important to continued stability that economic diversification continue. Every opportunity should be explored to create new jobs for these workers. Work such as that undertaken by the Indonesian Environmental Bamboo Foundation is important in identifying new markets and training workers in the use of new materials. These efforts should be supported in the future.

5. THE U.S. NEEDS TO BE FAR MORE ACTIVE IN ITS SUPPORT AND ENCOURAGEMENT FOR U.S. TRADE MISSIONS TO INDONESIA.

The delegation believes there is an enormous commercial opportunity for American companies in Indonesia. Projected imports for infrastructure-related projects in Indonesia total $113 billion over the next decade. The U.S. should be able to increase its exports to Indonesia significantly.

Many companies, particularly small and medium sized companies, are either unaware of these opportunities or unable to take advantage of them because they have no permanent representative in Indonesia. It is important that the U.S. government support and encourage trade delegations to Indonesia to increase awareness and support for these companies in identifying opportunities and learning to bid for new contracts. Such delegations should be coordinated with visits by high level officials to highlight the political importance the U.S. Government accords better trade linkages between our two nations.

V. COMMONWEALTH OF THE NORTHERN MARIANA ISLANDS

Visit

The delegation visited Saipan, the largest island and the commercial and governmental center of the Commonwealth of the Northern Mariana Islands (CNMI). Formerly a part of the Trust Territory of the Pacific Islands, which was administered by the United States under the terms of a United Nations Trusteeship Agreement, the CNMI entered into a permanent relationship with the United States in 1976 after approval of the Covenant embodied in Public Law 94-241. The Covenant was fully implemented in 1986 with the signing of Presidential Proclamation No. 5564, which terminated the Trusteeship Agreement as it applied to the CNMI.

In January 1978, the first constitutional government of the CNMI took office. While under the sovereignty of the United States, the people of the CNMI have enjoyed full local self government since that time.

The delegation believed it especially important to visit Saipan in 1994 to commemorate the fiftieth anniversary of the World War II battle for Saipan. The Empire of Japan had fortified the island in contravention of the League of Nations Mandate and used it as a base for the invasion of Guam at the opening of the War.

This year also began with the inauguration of the CNMI's fourth elected governor, the Honorable Froilan Tenorio, who assumed office on January 10, 1994, marking the third peaceful transfer of power after free and democratic elections.

These events are significant and important. They symbolize the transformation of the relationship between the U.S. and the CNMI from one of dependence to one of partnership. Both the CNMI and the U.S. have benefitted from this maturing relationship. The delegation believes it is important that this mutually beneficial foundation be remembered as discussions continue in this Congress on a number of important issues which have been the subject of heated and emotional debate in the past.

While on Saipan, the delegation had an opportunity to meet with Governor Tenorio and Lieutenant Governor Borja as well as other key past and present leaders from the executive, legislative and judicial branches of government.

Recommendations

1. AMERICAN WAR MEMORIAL PARK SHOULD BE COMPLETED THIS YEAR.

The delegation was encouraged by the Governor's pledge to take all necessary steps to complete agreements necessary to assure that the American War Memorial Park will be finished this year. The Park was established by the Covenant to commemorate the thousands of American GIs who served in the Marianas campaign and were willing to put their lives on the line for the people of Saipan in 1944.

2. STEPS SHOULD BE TAKEN TO RESOLVE PROBLEMS WITH CERTAIN NON-RESIDENT WORKERS.

The delegation was encouraged by pledges from Governor Tenorio that appropriate steps would be taken by the local government to resolve problems associated with nonresident workers employed in the garment industry in the CNMI. The delegation is convinced that it is critical that substantive progress be made as expeditiously as possible in this area if further divisive debate is to be avoided.

3. CONGRESS SHOULD CONTINUE TO ASSIST THE CNMI IN ADDRESSING INFRASTRUCTURE NEEDS.

The delegation took special note of the enormous economic growth in the Northern Marianas since 1976 and was impressed by the significant improvements that have been made in health, education and the general standard of living since then. The delegation also noted the continuing infrastructure needs of the CNMI, which continues to rely in part on pre-World War II systems.

4. GOVERNOR TENORIO'S INAUGURATION ADDRESS SHOULD BE THE BASIS FOR FUTURE COOPERATION IN ADDRESSING PROBLEMS IN THE CNMI.

The remarks of Governor Tenorio delivered at his inauguration prior to the arrival of the delegation contain a clear and, the delegation felt, hopeful statement of the task that lies ahead in the CNMI. Members of the delegation pledge to work with the governor in the spirit of these remarks to see that these goals are achieved.

VIEWS OF SENATOR ARLEN SPECTER

I am pleased to offer my thoughts on the January 1994 Congressional Visit to China, Vietnam, Thailand and Indonesia as part of the report prepared by Senator Johnston on behalf of the delegation.

Our visit to Asia provided an opportunity for the members of the delegation to understand the potential of two-way trade with the Asian-Pacific Rim. Trade with these countries already exceeds our commerce with the European community or a combination of our North and South American neighbors.

The potential for increased trade and more U.S. jobs springs from the fact that capitalism is on the march in Asia. If Adam Smith and Karl Marx could visit Beijing and Hanoi today, they would be shocked to see countries like Red China and Communist Vietnam sprinting toward capitalism.

Officially they call it a "socialist market economy," an oxymoron which may be the practical answer for countries not wanting to admit their wish to abandon the disadvantages of a planned economy for the rewards of individual initiative. On the same day in early January when the delegation heard China's Premier Li Peng in Beijing extol the theoretical virtues of socialism, the Vice-Chairman of The National People's Congress announced plans to copy economic legislation from capitalist countries.

The delegation heard a similar message on our visit to Hanoi. A young official from the Vietnam Ministry of Trade was blunt and succinct in stating that the private sector was more effective than state ownership because "people work harder for their own pockets." Looking to western political ideology, Vietnam's most powerful political figure, Secretary General Do Muoi of the Communist Party, claimed his country sought to follow Lincoln's model of a government of the people, by the people and for the people.

Statistics show China's gross national product growing at about 10 percent a year. Premier Li Peng boasted of a trillion dollar Chinese market in urging our delegation to renew most favored nation status for China. Similarly in Thailand, Indonesia and Vietnam, we heard of multi-billion dollar U.S. opportunities in hydro-power, telecommunications, air and rail transport, automobiles and environmental projects.

In all these countries, U.S. businessmen urged our delegation to delink trade interests from human rights and even the MIA issue. Frank Hawke, representing CitiBank's global finance operation in Asia, pointed out that about half of his company's $2,000,000,000 annual profits came from developing countries. Members of the American Chamber of Commerce in Beijing chided our policy of lecturing China on human rights when 23,000 people were murdered annually in the U.S. and our cities were virtual war zones.

When we objected to China's missile sales, Premier Li Peng chastised us for being the world's number one "big brother" in arms sales. Indonesian officials expressed concern about continuation of their trade privileges under General System of Preferences when our delegation raised the issue of human rights in East Timor and labor conditions generally.

Those protests suggest that our conditions for trade preferences are having an effect. Access to our markets at reduced tariff rates are vital to Beijing's and Jakarta's trade so we should continue our leverage on human rights and arms sales without breaking the lever. It is obviously a judgment call as to how hard to press; but there is much room for improvement and we are having significant success as evidenced by China's recent concession to allow greater inspection of manufacturing in their prisons.

Regrettably, a tough trade policy and sanctions also seem necessary to enforce U.S. property rights. China has flooded our markets with illegal textile trans-shipments. After cutting their import quotas, we have finally extracted commitments from China to stop the illegal trans shipments of textiles, but that will have to be monitored closely to assure compliance.

Responding to longstanding U.S. complaints, Thailand is finally enacting tough legislation and establishing a special court for enforcing intellectual property rights. Similarly, Indonesia needs continuing pressure and monitoring to fulfill existing promises to respect U.S. copyrights.

Notwithstanding the problems, the big picture suggests enormous U.S. opportunities in the Pacific Rim. China's Deputy Power Bureau Director Yu Fu Min was outspoken in his preference for U.S. products over France and other competitors. Within the next decade, Indonesia will provide a market for infrastructure development in excess of $100 billion.

The potential for U.S. business development in Indonesia is illustrated by a multibillion dollar copper-gold mine being developed by Freeport-McMoRan Inc. in Timika in East Indonesia. In the past five years, that talented and ambitious company has developed the world's third largest copper mine creating thousands of new jobs there and in the United States.

With respect to the lifting of the trade embargo with Vietnam, it is imperative that the United States do everything possible to fully account for the 2,239 servicemen missing and unaccounted for in Vietnam. During the trip to Vietnam, the delegation had the opportunity to meet with Major Gen. Thomas Needham, who is in charge of the U.S. military efforts on the MIA issue, and the other U.S. military personnel in Vietnam. We also spoke with people from the Vietnamese Government who are trying to cooperate in producing the remains of all the U.S. servicemen.

I share the conclusions articulated by Senator Johnston and others on the delegation that it appears that the Vietnamese Government is trying to cooperate in the accounting effort. However, I do not know if the actions by the Vietnamese represent the maximum good-faith effort possible. During the trip I made it clear that I could not join my colleagues on the delegation in seeking removal of the trade embargo until I was more confident that every effort possible is made to resolve the MIA issue.

On my return to the United States, I spoke with Senator Bob Smith who has argued that the Vietnamese have not been fully cooperative. He has been careful to back up his generalizations with specific evidence, that the Government of Vietnam can do more to disclose the location of the remains of U.S. service personnel.

I believe the President of the United States is in the best position to make the determination to end the embargo. The President, with his executive authority and access to much more information is best equipped to make this decision.

The question which must be addressed is whether the Vietnamese Government has made a full accounting and whether there has been full disclosure of facts on the MIA's. Given the information available to me, it appears that there has not been complete disclosure and cooperation.

The President on February 3, 1994, announced the lifting of the embargo. Clearly the United States has an interest in improved U.S.-Vietnam relations. Closer trade relations with Vietnam will allow U.S. companies access to the developing Vietnamese economy. However, we should not concede our foremost objective—to gain greater accountability of the missing servicemen. It is my hope that the Administration will use its maximum remaining leverage to ensure that the Vietnamese Government provide honest assistance in producing the remains of the U.S. servicemen.

ACKNOWLEDGMENTS

This report was prepared with support from the Global Studies Program at Battelle, Pacific Northwest Laboratory; the U.S. Environmental Protection Agency; and the U.S. Department of Energy. The author would like to thank William U. Chandler, Meredydd Evans, Sonny Kim, Elizabeth Malone, and John W. Parker of Battelle, Pacific Northwest Laboratory for editing and advice.

LINKING ENERGY AND ENVIRONMENTAL POLICY IN CHINA

China's surge in energy demand may become an environmental disaster extending well beyond its borders unless steps are taken to reduce the country's inefficient use of energy and its heavy reliance on coal. This paper explores:

The extent of the environmental damage inflicted by unrestrained growth of energy demand, including respiratory disease, acid rain, and the threat of global warming;

Chinese policies and programs to reduce the impacts of coal use by promoting cleaner and more efficient energy use and the use alternative fuels; and

International cooperation between China and the U.S. government, multilateral development banks, nongovernmental organizations, and private firms to promote sustainable energy use.

I. Trends in coal and oil use in China

China is the world's largest producer and consumer of coal, and its plentiful reserves are expected to last for at least another 100 years. Coal is the fossil fuel that, along with oil, has fired China's explosive economic growth—almost nine percent per year on average since the introduction of market reforms and decentralization in 1978.

Coal is the only fossil fuel in China that is truly abundant on a per capita basis. China had 719 tons of proven reserves of coal per capita, compared with only 5.2 tons of proven reserves of oil per capita in 1987.[1] Coal accounted for 74 percent of China's commercial energy use in 1991, while oil accounted for 19 percent. China relied on hydropower for five percent of its energy, and natural gas accounted for only 2 percent.

The Chinese not only burn large quantities of fossil fuel, but they burn it in inefficient and highly polluting ways. Therefore, despite success in limiting growth of energy use to half of GDP growth during the 1980s, fossil fuel use has contributed to the high environmental price China has paid for its booming economy.

China has nearly doubled its coal production from 620 million tons in 1978[2] to 1.1 billion tons of raw coal in 1992, and plans to boost production to 1.4 billion tons by 2000.[3] Oil production grew more slowly than coal during the reform period, from 100 million tons in 1978[4] to 140 million tons of crude in 1991.[5] Oil consumption actually declined in the early 1980s as China pursued a policy of "Replacing Oil with Coal", i.e., exporting oil to earn foreign currency to buy more efficient foreign coal-fired power plant equipment. In contrast to oil, most of China's coal is consumed domestically.

The residential sector

Chinese per capita consumption of commercial energy is less than one twelfth that of the United States, and Chinese households consume 36 times less energy and 75 times less electricity than their U.S. counterparts.[6] China's level of residential electricity service is low, but it is growing quickly. With the additional income generated by economic reforms, many Chinese are rushing out to buy refrigerators, air conditioners, washing machines, and other appliances. China produced an estimated 6.8 million refrigerators in 1992, 18 percent more than during the previous year. The share of electricity used by households almost tripled from 3 percent in 1980 to 8 percent in 1991.[7]

Yet basic needs such as heating and cooking continue to absorb most of residential and commercial energy use.[8] Central heating is often unavailable and electric heaters and cookware are usually prohibited, so most families heat their homes and cook their food with coal stoves. In fact, coal, often unprocessed, accounts for 80 percent of domestic energy use,[9] not including biomass. Coal consumption is constrained by price and availability, not by demand. As family incomes rise and coal transportation improves, this sector has tremendous potential for growth.[10]

The industrial sector

While residential use is on the rise, industry continues to consume the lion's share of China's energy, with the nonstate sector leading the way. Almost 70 percent of China's energy was used by industry in 1991. The nonstate sector, which grew at an annual rate of 17.6 percent during the 1980s, is primarily responsible for China's industrial revolution and corresponding increase in overall energy demand. The nonstate share of total industrial output has expanded from 22 percent in 1978 to 45 percent in 1990.[11]

Only a small proportion of nonstate enterprises—individual enterprises and joint ventures—are privately owned and operated. The majority are urban collectives or town and village enterprises (also known as TVEs or rural enterprises), which are owned and run by local collectives and are closely related to local governments.

TVEs expanded production and profits at the extraordinary rate of 50 percent last year, increasing their share of industrial output from 30 percent in 1992 to an estimated 40 percent in 1993.[12] Unfortunately, many of these new factories use inefficient, polluting, second-hand equipment and are supplied by small, inefficient power plants.[13] TVEs also represent an inherent conflict of interest in terms of energy efficiency and environmental protection because the same officials who run the factories often have the additional responsibility of ensuring that the plants comply with regulations.

Although environmental awareness is growing, most top political leaders still put production quotas before environmental protection. They would like to use cleaner technologies but feel that China is too poor to sacrifice economic growth for environmental protection.

II. ENVIRONMENTAL IMPACTS OF COAL AND OIL USE

Dirty, inefficient fossil fuel use is widely recognized as one of China's most pressing environmental problems. A recent survey of environmental experts and scientists listed the development of clean and highly efficient coal combustion technologies as a top environmental priority for China, second only to developing cleaner production processes in pollution intensive industries.[14] The most harmful pollutants associated with coal combustion are particulate emissions and sulfur emissions. Most air quality protection efforts have focussed on controlling these two pollutants, with limited success.[15] Carbon dioxide emissions and the corresponding threat of global warming also pose great dangers to China's environment.

Particulates and respiratory disease

Emissions

High levels of particulate emissions are cause for concern primarily because they damage the human respiratory system. The State Statistics Bureau reports that annual emissions of particulates from enterprises alone in the 1980s ranged from 13 to 15 million tons. Although a large share of environmental protection funds was reportedly used to control them, particulate emissions data do not show a downward trend during this period. Pollution control efforts seem to have at least prevented the situation from getting worse as coal combustion steadily increased.[16] The National Environmental Protection Agency reports that, as of 1991, China was emitting 23 million tons of particulates annually, with coal burning accounting for about 17 million tons.[17] The National Environmental Protection Agency number is higher than the State Statistics Bureau number because it includes residential use. China may be burning 1.4 billion tons of coal per year by 2000. At current controls, that means that annual dust emissions will climb from 23 to over 30 million tons.[18]

Changes in particulate emissions levels are an important way of measuring technological progress or lack thereof, but examining the concentration of total suspended particulates (TSP) is the key to determining whether control efforts are reducing harm to human health by improving ambient air quality. TSP data are sketchy—only a few cities have complete records for the decade, and those that do, like Shenyang in Liaoning Province, were usually the sites of concerted efforts to reduce pollution from an extremely high level and therefore are probably not typical of the nation as a whole. These data also hide the fact that many pollution sources may have been relocated to rural areas rather than actually eliminated.

Ambient air quality is generally better in the south than in the north because of variations in the amount of coal burned for heating in the winter. TSP concentrations vary greatly from city to city. The large, prosperous coastal city of Guangzhou, for example, was at the low end of the scale at 280 micrograms per cubic meter, while Taiyuan, at the heart of China's coal mining region, averaged 780 micrograms per cubic meter.[19] Those cities that made the effort seem to have reduced TSP levels considerably, but this conclusion may be premature because annual variation due to weather is considerable.[20]

Respiratory problems

Respiratory problems are very common in northern cities during the heating season. The question is not whether coal smoke damages the lungs, but how much different factors contribute to respiratory disease. Studies done in Xuan Wei County in Yunnan Province and several cities in Liaoning Province assess the independent and synergistic effects of pollution, smoking, and coal burning in the home.

Joint U.S.-Chinese research in Xuan Wei County established a clear link between lung cancer and the use of unvented coal pits in houses for heating and cooking. Overall, death from lung cancer is still much less common in the China than in the United States. Statistics age-adjusted to China's population in 1964 show that 4 times fewer Chinese males died of lung cancer and half as many Chinese females died of cancer than in the United States. In Xuan Wei County, however, lung cancer mortality rates among females are four times higher than in the United States. Female lung cancer mortality rates in communes burning "smoky" coal in Xuan Wei County are 20 times higher than in the United States.[21] The Xuan Wei study heiped Chinese scientists to publicize the health advantages of using stoves and chimneys that are better ventilated, and the results of the study may also be used to develop environmental standards.[22]

Respiratory disease studies were also done in Liaoning Province,[23] located in Northeast China, the home of China's heavy industry. Liaoning's capital, Shenyang, is one of China's most polluted cities. Studies done in cities in Liaoning show mortality rates from chronic obstructive pulmonary diseases of 80 per 100,000, 4 times higher than rates in developed countries. In Shenyang, age-adjusted mortality from lung cancer is around 50 per 100,000 for males. The rate for females, 29 per 100,000, is among the highest in the world. Air pollution, both indoor and outdoor, accounts for an estimated 60 percent of chronic obstructive pulmonary diseases and 25 percent of lung cancer. Every year in the Shenyang urban area, an additional 300,000 people are treated and an additional 2000 people die of chronic obstructive pulmonary diseases due to all pollution. The cost in medical expenditures, lost working days, and loss of life before retirement age is $31 million.[24]

A study of 703 pairs of cases of lung cancer and controls conducted in 7 cities in Liaoning found that cigarette smoking is still the principal cause of lung cancer, accounting for 55 percent of the cases in men and 37 percent of the cases in women. By contrast, smoking is responsible for 85 percent of lung cancer cases in developed countries. Air pollution was the second most significant factor in China, increasing the risk of lung cancer by 50-60 percent.

Another Liaoning study examined the effects of pollution, coal burning, and smoking on the prevalence of bronchitis. Researchers were able to examine independent and synergistic effects of these factors by comparing groups of people exposed to one or more of the factors to a control group composed of nonsmokers living in relatively clean areas who do not burn coal in their homes. In terms of independent effects, heavy pollution was the most serious, followed by smoking, moderate pollution, and coal burning. The effects of heavy pollution can be seen most clearly by noting that the rate of chronic bronchitis among residents of areas of heavy pollution who neither smoked nor burned coal was more than 10 times higher than that of the control group. Those who also burned coal in their homes were more than 12 times as likely as the control group to have bronchitis. Nonsmokers in areas of moderate pollution had 3 times the rate of bronchitis as the control group if they did not burn coal in their homes, and 5 times the rate of the control if they did.

Sulfur dioxide and acid rain

Emissions

Sulfur dioxide is a precursor to acid rain. Sulfur dioxide emissions from industry declined from 16 million tons to 12 million tons between 1980 and 1986, but rose again to 16 million tons by 1989. Given that coal use grew steadily, installation of scrubbers and other equipment may have caused the reduction of emission before 1986, while the subsequent rise may reflect a tapering off of emission control improvements as the incremental cost of better control equipment increased.[25] Sulfur dioxide emissions may rise from 15 to 20 million tons by the year 2000 as coal use continues to grow.[26] The Chinese government initially announced that they intended to cap sulfur dioxide emissions at 20 million tons by 2000,[27] but the target has been raised to 21 million and then to 21-23 million. China would have to spend $2.6 billion, or $520 million annually, to reduce sulfur dioxide emissions by the 2 million tons required to meet the year 2000 target.[28] While far from insignificant, this amount is small compared to the cost of a new power plant, up to $3.5 billion.

China produced 17 million tons of sulfur dioxide in 1992 according to official figures, but the actual total was closer to 19 million tons if estimated emissions from town and village enterprises are included. The top two sectors producing sulfur dioxide in China are power production and chemical industries. The three main types of equipment producing the most sulfur dioxide are industrial boilers, power plant boilers, and household stoves. China has put less effort into controlling sulfur dioxide emissions than into reducing particulate emissions, and did not achieve similar success in reducing sulfur dioxide emissions because sulfur dioxide control technologies are much more expensive.

Ambient concentrations of sulfur dioxide in cities did not rise as fast as coal consumption, however, because of environmental protection measures and higher smokestacks. The government has begun collecting a sulfur dioxide emission fee in two provinces and nine cities. The funds will be used to purchase sulfur control equipment.

Acid rain impacts

China first began to study acid rain in 1982 and conducted in-depth investigations from 1986 to 1988. Researchers discovered that although there was more pollution in the north, acid deposition was more severe in the south, probably because northern alkaline aerosols in the north are able to neutralize the acidity. Acid rain is particularly severe in South China and the Sichuan Basin. Acid rain causes significant agricultural losses, but the most serious effect is on forests.[29]

Sulfur dioxide emissions create precipitation below a pH of 5 in an area greater than 1.3 million square kilometers, around 14 percent of China's total land area.[30] Much of the damage from acid rain occurs in southwest China, where coal tends to have a high sulfur content (around 3 to 6 percent).[31] In addition much of the coal is unwashed, and is burned in small and medium- sized inefficient industrial boilers and household stoves.

Guizhou Province in southwest China suffers from serious acid rain problems. As of 1991, acid rain covered about two thirds of the total area in central Guizhou, the pH of precipitation is below 4.5, and total sulfur deposition is about 2.5 tons per square km per year. Precipitation pH is around 4 in most cities and sulfur deposition reaches 10 tons per square km per year. The lifetime of zinc-plated steel, normally 29 years, is reduced to 4 years in the capital city of Guiyang and 9 years in rural Guizhou. Economic losses due to damage to metal, building materials, forests, and crops are estimated at over US$100 million.[32]

Researchers have recently recorded pH values of rain as low as 2.6 in Guangxi Autonomous Region. The annual average pH of rainwater was 4.5 in 1992. The rate of acid rain in different areas ranged from 13 to 88 percent, higher than in 1991. The concentrations of sulfur dioxide in ambient air were lower than in 1991, but still exceeded national guidelines by up to 2.6 times in major cities like Nanning and Liuzhou.[33]

South China is the third largest area in the world seriously affected by acid rain, after Europe and North America. Compared with those areas, China's acid rain is not very serious, but the trend toward ever-greater sulfur dioxide emissions is cause for concern for both China and its neighbors, as high smokestacks spread acid rain farther from industrial sources. Direct losses, such as current damage to forests, can be calculated, but indirect and incalculable losses, such as soil deterioration and damage to ancient buildings and statues will perhaps have a greater impact in the long run.

Carbon dioxide emissions and climate change

Emissions

China produced 580 million tons of carbon by burning fossil fuels in 1990. Carbon emissions are expected to grow by 1-1.5 billion tons by 2025, making China the world's largest emitter.[34]

Climate change impacts

The risk of global climate change and its potential impacts on oceans, agriculture, and urban areas are very controversial topics. A warming trend has already been detected in China, but whether it is due to global climate change is uncertain. Average temperatures over China have risen by 1.0 °C since the 1850s—faster than the global average of 0.5 °C. A report prepared by a team of researchers from China, the United Kingdom, the Netherlands, and Switzerland predicted additional warming above the 1990 level of 1.2 °C by 2050 and 2.5 °C by 2100, a rate five times that of the previous 100 years.[35] Corresponding impacts on China's climate would also be slightly more rapid than for the world as a whole. Temperature rise is expected to produce an overall decrease in soil moisture, reducing the areas of alpine vegetation in the mountains of southwestern China, and transforming the temperate desert and steppe of Xinjiang Province into warm or subtropical desert. Effects on agriculture are expected to be mixed. Rice production is expected to fall because although the area with temperatures suitable for growing rice will increase, water will be more scarce. Warming in northeast China is expected to lead to higher grain yields.[36] Another modelling study assumed warming of 1-4 °C and precipitation variations of 10-20 percent. The model showed that under current agriculture practices, corn yields would suffer the most, followed by wheat, and then rice. Agricultural conditions will improve in northern China, remain the same south of the middle and lower reaches of the Yangzi Kiver, and deteriorate in central China.[37]

The most direct impact of climate change on people's lives may be rising sea levels. Records from the past 20 to 30 years indicate that the average rate of sea level rise for the East China Seas was 0.14 centimeters per year, approximately equal to the global average. Sea level rise has already begun to increase beach erosion and salinization of coastal lowlands and plains, intensify the effects of coastal hurricanes and typhoons, and increase the vulnerability of estuarine areas to floods. Most of China's coastal areas have retreated, with the exception of river deltas that are constantly replenished with sediment from erosion upstream. The effects of erosion can be seen clearly in Jiangsu Province. Since the Yellow River changed its course in 1855 and its mouth shifted from Jiangsu to Shandong Province, the abandoned mouth in Jiangsu has retreated more than 20 kilometers, and as much as 1400 square kilometers of adjoining land have eroded away. In addition to eating away at the shoreline, seawater also seeps into groundwater. Seawater invasion is particularly severe in the coastal plain to the south of the Bohai Sea due to light precipitation and excessive groundwater pumping—seawater has salinized an area of 430 square kilometers along 200 kilometers of coastline in that area. Higher sea levels also raise river water levels. This increases river bank erosion which raises the height of river beds, thus greatly increasing the risk of flooding, especially in the lower reaches of the Yellow River.[38]

Future increases in sea level may have even more dire consequences. Vulnerable coastal areas include the Yangzi, Pearl, and Yellow River deltas that have population densities of more than 800 per square kilometer and produce a significant proportion of China's GNP. Global sea level is expected to rise by between 15 and 90 centimeters. The mean elevation of four of China's major cities—Shanghai, Tianjin, Guangzhou, and Shenyang—is between only 2.2 and 6.6 meters above current sea level. The problem is exacerbated by the current tendency of land to sink along the coast. An extreme case is Tianjin, which has sunk by about 50 centimeters relative to current sea level over the last 30 years. Sea level rise of 50 centimeters would flood 40,000 square kilometers of the densely populated Liaohe, Haihe, and Subei Plains, potentially forcing millions of people to migrate to higher ground.[39]

Other energy-related pollution

Particulates, sulfur dioxide, and carbon dioxide emissions are the main, but not the only, environmental pollutants produced by the energy industry. Others include the greenhouse gas methane from coal mines, smog-inducing oxides of nitrogen and volatile organic compounds from vehicle exhaust, and various water pollutants and solid waste. Several cubic meters of methane are released for every ton of coal produced, i.e., several billion cubic meters in 1992. Only a small portion of coal bed methane is used as fuel, the rest being released to the atmosphere and contributing to the greenhouse effect.[40] NO_3 concentrations for seven Chinese cities (Beijing, Tianjin, Langfang, Tanggu, Jixian, Guangzhou, and Liuzhou) average 19.5 micrograms per cubic meter, dwarfing Los Angeles's 6 micrograms per cubic meter, although it is unclear what proportion of NO_3 comes from fuel combustion.[41] Energy production and use creates vast quantities of waste water, much of which is dumped untreated into neighboring rivers, streams, and lakes. Water is wasted and polluted when coal is mined and washed, and when oil is recovered and refined. In 1989, 1.5 billion tons of mostly unused ground water were discharged from coal mines.[42] As of the same year, only about half of all enterprises with coal washing plants had a closed system that treated and reused waste water. The remainder discharged untreated water to the environment, a total of 45 million tons.[43]

Spontaneous combustion of mining waste poses safety as well as environmental hazards. Two billion tons of worthless rocks have accumulated as a result of coal mining, with 100 million more tons added every year.[44] The waste rock can also block rivers. For example, the Environmental Protection Committee of the National People's Congress recently reported that mining activities threaten to clog the Wulanmulun River in the Shanxi-Shaanxi-Inner Mongolia border region with waste, increasing the risk of flooding.[45] The problem is exacerbated by the irresponsible behavior of large numbers of unregistered private miners. Fires in both coal and oil fields are serious problems. Mining also causes subsidence, the sinking of land located above coal mines. Subsidence damages overlying buildings and railroads, and reduces agricultural output by destroying land and irrigation projects. One thousand square kilometers of land have been destroyed by subsidence.[46]

III. ENERGY AND ENVIRONMENTAL POLICY IN CHINA

The Chinese government is well aware of the aforementioned environmental ills and has taken steps to remedy them. Because funding is limited, solutions that combine environmental protection with economic development, such as energy efficiency, are the most likely to receive government support. However, improving energy efficiency is not a sufficient means to control energy pollution. The Chinese government also recognizes the importance of switching from coal to renewables, natural gas, and oil, and developing cleaner methods of producing and burning coal and oil.

Background

A brief comparison of energy management systems in the United States and China provides the context in which China's energy and environmental policy is conducted. In the United States, fossil fuel is generally produced by private companies and sold at market prices. Electricity is produced and distributed by private, municipal, state, and federal utilities and sold at prices set by state public utility commissions. The U.S. Department of Energy implements energy laws and regulates private firms; for example, setting and enforcing energy efficiency standards for appliances or designing regulations that give tax incentives for investing in efficiency.

China's energy sector, on the other hand, is now in an awkward transitional stage between central planning and a private enterprise system. Energy ministries are becoming energy corporations, but they retain some government functions and no U.S. Department of Energy-type organization has emerged to regulate them. The state requires state, collective, and private fossil fuel producers to sell a certain amount of their product to the state at a state-set price called the "low plan" price. Enterprises can sell amounts above the quota to the state at the high plan price or sell it on the market. Fuel bought by the state is sold at low prices to state enterprises. New power projects are owned by both public and private investors, but prices are chaotic and irrational, and do not encourage efficiency or load balancing. Energy efficiency programs are designed by the State Planning Commission and efficiency investments are made by the State Energy Investment Corporation. The state is attempting to shift responsibility for energy efficiency investments from itself to enterprises by creating regulatory and market incentives.

CHINA'S ENERGY EFFICIENCY POLICIES

Recognizing that efficiency improvements were essential to meeting its ambitious targets for economic growth, the Chinese government launched a multi-pronged attack on energy waste beginning in the early 1980s. Its efforts were not in vain: a combination of macroeconomic restructuring and technological renovation allowed China's economy to grow twice as fast as energy consumption, an unprecedented feat for a developing country.

Price reform

China is plagued by energy shortages. In a free market, energy shortages would result in high energy prices, creating an economic incentive to improve demand side efficiency and bringing supply and demand into balance. In China, however, energy prices have historically been kept low and, despite significant price hikes in the last 15 years, they still average below the international market price. Since the late 1970s, Chinese energy prices have often failed to cover the cost of production, forcing state coal mines and other energy enterprises to rely on heavy subsidies from the central government.

As part of the economic reform program the government has given state enterprises an ultimatum: profit or perish. The state will stop paying the coal industry its annual subsidy of $230 million. To make ends meet, China General Coal Corp., which produces a third of China's coal, raised the percentage of its output sold at market prices (more than twice the plan price) to 57% in 1993 compared to only 20% in 1992.[47] All coal has been sold at market prices since 1 January 1994. Oil and electricity prices have also been rising, but recent measures have slowed down the process.[48] The government is concerned that accelerating energy price reform could cause social unrest because industry is likely to pass on the higher energy costs by increasing prices of consumer goods.

Institutional reform

Price reform is only half the battle. China's efficiency drive also involves promoting competition and accountability by freeing energy enterprises from state ownership and control. Under traditional central planning, all enterprises were owned and operated in one of two ways: state ownership (central government) or collective ownership (provincial government and below). Five Year Plans drawn up by the State Planning Commission determined production and allocation of goods from both types of enterprises. Production ministries, such as the Ministry of Coal and the Ministry of Oil and Gas, implemented the plans. Enterprises turned over all profits to the central government and the state subsidized enterprises' losses. Private enterprise was not allowed. Lack of accountability and the absence of a profit motive dampened the incentive to improve energy efficiency.

China has taken some halting steps toward privatization. Rather than selling off state assets to the private sector, China has turned some energy ministries, or parts of ministries, into state energy corporations. Some observers applaud the success of the state-owned Huaneng Group which built one third of China's power plants in the 1980s with seed money from oil exports, but critics contend that Huaneng is still financially linked to the central government and therefore is not truly accountable. These critics recommend increasing competition by breaking up Huaneng and other large energy corporations like Sinopec, which holds a monopoly in the oil products retail sector.

The emergence of state corporations has been accompanied by the rise of nonstate enterprises. The state's share of production and consumption is declining rapidly, but collective ownership persists, with only a small portion of GDP coming from the private sector. In 1993, 45 percent of China's production and profits came from the nonstate sector, mostly town and village collectives, a slightly different breed from the private businesses officially known as "individual enterprises" (8 employees or less) and "private enterprises" (more than 8 employees). Town and village enterprises (TVEs) are common in the coal industry because coal, unlike oil, does not require sophisticated equipment to retrieve. While the state continues to own and operate large coal mines and gives preference in coal allocation to large state-owned heavy industrial enterprises, such as iron and steel manufacturers, local mines have jumped in to meet the growing coal demand in the nonstate sector.

Decentralization has also occurred in the power sector. Local governments capable of raising sufficient funding, including foreign exchange to import equipment, usually have no trouble getting approval from the central government to build new power plants. Large private enterprises are also allowed to invest in plants that will provide them with power. The new power plants are accountable for profits and losses, and are allowed to charge higher tariff rates than existing plants, but the final price must still be approved by the Ministry of Power.[49]

State investment in efficiency

Direct investment in efficiency as well as macroeconomic reform helped China decouple energy use and economic growth in the 1980s. The Chinese government invested significant resources in improving efficiency in key energy-intensive industries, as well as in improving the efficiency of motors, fans, and pumps in all industries. However, total state investment in demand-side measures was dwarfed by investment in new power plants. The central government invested approximately $3.8 billion in expanding power supply last year,[50] about ten times what it spent on energy conservation.

There are two main channels for direct state investment in energy conservation: the Energy Conservation Investment Company (under the State Planning Commission) which is responsible for large scale projects, and the Energy Conservation Technological Renovation Fund (under the State Economic and Trade Commission) which covers small scale projects. The two organizations changed the form of investment from grants to loans in 1985, but interest rates for energy conservation loans are 50 percent lower than the prevailing rate. The standard rate for energy conservation loans in 1993 was 4.8 percent. Interest rates may be reduced by even more than 50 percent for projects with significant environmental and/or social benefits.[51]

No loan guarantees are required because government agencies administer the loans. In the first step of the process, the enterprise submits a technical proposal for renovating equipment or improving processes to the provincial or municipal energy conservation office. This office then selects the best proposals, about one third of the total, and submits them to either the Energy Conservation Investment Company or the Conservation Fund, depending on the scale of the project. The State Planning Commission and the Ministry of Finance determine the total amount of funding available for energy conservation loans that year and how the total will be divided among the local energy conservation offices and ministries. In effect, the State Planning Commission and the Ministry of Finance give the subordinate agencies permission to borrow a certain amount of money from a state bank, the Industrial and Commercial Bank in most cases, or the Bank of China if foreign currency is needed. The energy conservation offices then distribute loans to the enterprises.[52]

The Energy Conservation Investment Company's funds have grown by 20 percent annually since 1990. The company invested $300 million in 1993. The Conservation Fund administers $70-87 million in loans per year. Central government loans are supplemented by funds from local governments and enterprises. Overall local governments invest approximately 40 percent as much as the central government in energy conservation projects. The share contributed by enterprises varies. In Beijing, for example, enterprises must contribute half of the investment.[53]

Investment was focused in the following areas in the 1980s:

- Cogeneration;
- Replacing small generators with larger ones;
- Fertilizer production;
- District heating;
- Iron and steel recycling;
- Continuous casting;
- Small cement kiln renovation;
- Coal washing and blending; and
- Use of waste heat to generate electricity.

Small-scale projects included waste heat recovery and use in space heating, small cogeneration, and boiler renovation.

China's energy conservation management system will undergo major changes as China decentralizes and shifts emphasis from planning to markets. The government is expected to create more incentives for saving energy, such as emissions trading programs, and penalties for wasting it, such as fines for exceeding emissions limits or using inefficient equipment. Government agencies will probably require banks to set aside funds for conservation, but will allow them to fund specific projects based on commercial feasibility.[54] Energy conservation policy will also need to expand to include TVEs, which received no assistance in the 1980s. In fact, much the old equipment "eliminated" by state and local firms was absorbed by the booming rural industrial sector. In the 1990s, the Rural Energy Department of the Ministry of Agriculture provided some funds for TVE energy conservation, but they are not sufficient. To remedy this problem, lenders will be encouraged to evaluate projects based on technical and commercial feasibility and environmental and social benefits, not based on who operates the firm, thus giving TVEs equal access to energy conservation loans.[55]

Energy pollution control policy

Policy tools

Most energy pollution is air pollution. China has several general policies that affect air pollution control, as well as measures specifically aimed at energy pollution, such as coal smoke and vehicle emissions control. (See Table 1.)

Table 1.—AIR POLLUTION LAWS, REGULATIONS, AND STANDARDS PROMULGATED FROM 1980–89[59]

Name	Issuing agency
Air Pollution Prevention and Control Law of the PRC.	Standing Committee of the National People's Congress
Air Pollution Prevention and Control Measures	National Environmental Protection Agency
Regulations Regarding Coal Smoke Prevention and Control Technologies and Policies.	Environmental Protection Committee of the State Council
Method for Managing Air Pollution Control Regions in Cities.	Environmental Protection Committee of the State Council
Method for Developing Residential Coal Briquettes.	Environmental Protection Committee of the Briquettes State Council, State Planning Commission, State Economic Commission, Ministry of Finance, State Pricing Bureau, State Commodities Bureau
Ambient Air Quality Standards	Environmental Protection Leadership Small Group of the State Council
Boiler Emissions Standards and Monitoring Method Standards.	Urban and Rural Construction Environmental Protection Department
Industrial Boiler Emissions Standards and Monitoring Method Standards.	National Environmental Protection Agency
Idling Vehicle Emission Standards and Monitoring Method Standards.	Urban and Rural Construction Environmental Protection Department
Diesel Vehicle Acceleration Exhaust Standards and Monitoring Method Standards.	Urban and Rural Construction Environmental Protection Department
Diesel Motor Full Load Exhaust Standards and Monitoring Method Standards.	Urban and Rural Construction Environmental Protection Department
Vehicle Crankcase Emission Standards	National Environmental Protection Agency
Light Vehicle Emission Standards and Monitoring Method Standards.	National Environmental Protection Agency

The government enforces emissions standards through fines, and through orders to curtail operations, relocate, or shut down. Article 18 of the Environmental Protection Law authorized the collection of discharge fees in 1979.[56] Discharge fees are set far below the cost of adopting pollution control technologies, even below the cost of operating pollution control equipment. Fees nevertheless serve the purpose of generating funds to cover environmental protection administrative costs and the cost of improving pollution monitoring. Environmental Protection Bureaus also use fees to establish loan funds to help firms pay for installing pollution control equipment. These loans (formerly grants) have accounted for approximately half of pollution control expenditures in China. The loan system has made relations between Environment Protection Bureaus and firms are less adversarial than in the United States.[57]

China has initiated a sulfur dioxide emission fee collection pilot project in the following provinces and cities: Guizhou, Guangdong, Chongqing, Yibin, Nanning, Guilin, Liuzhou, Yichang, Qingdao, Hangzhou, and Changsha. The fee is $0.03 per kilogram of sulfur dioxide.[58]

Technical measures

China has made concentrated efforts since the 1980s to reduce residential energy pollution by installing district heating, switching from coal to gas, and distributing coal briquettes. The replacement of household-based heating with central heating had, as of the end of 1989, reduced coal use by 1.5 million tons, reduced dust emissions by 73,000 tons, and reduced sulfur dioxide emissions by 23,000 tons. A goal of 470 million square meters has been set for district heating space. Replacing of coal with gaseous fuels for kitchen ranges saves 3.5 million tons of coal, reduces sulfur dioxide emissions by 59,000 tons, and reduces municipal waste by 1.05 million tons per year. Compared to raw coal combustion, briquette combustion reduces CO_2 emissions by 70 to 80 percent, reduces particulate and 3.4 benzopyrene emissions by 90 percent, and reduces sulfur dioxide emissions by 40 percent when lime or calcium carbide wastes are included in briquette binding material.[60]

These efforts have produced noticeable results. In Beijing, for example, the skies were clearer this winter (1993-94), although air pollution levels still exceeded national guidelines. More people have moved from courtyard homes with individual coal stoves to centrally heated apartment buildings with more efficient and potentially cleaner boilers, and the city government is building a pipeline in the western part of the city to supply natural gas to approximately one million households. Emissions from the growing numbers of motor vehicles that now clog the city's streets, however, threaten to erode these improvements in air quality. Environmental protection agencies will impose new vehicle emissions standards soon, but they will have trouble enforcing them.[61]

Technical measures to reduce energy pollution in the industrial sector cover all stages of the coal preparation and use. They include pre-combustion measures such as coal washing; adopting advanced clean combustion technologies (e.g., circulating fluidized bed combustion); and end-of-pipe solutions, such as electrostatic precipitators, sulfur scrubbers, effluent treatment, and fly ash utilization. Some measures, such as tall chimneys for power plant boilers, may reduce local pollution, but only at the cost of spreading emissions over a larger area.[62]

IV. CASE STUDIES IN INTERNATIONAL COOPERATION

China has encouraged foreign involvement in its efforts to combat energy pollution. Below are some examples of public, private, and nonprofit sector efforts to improve energy efficiency, reduce the impacts of coal use, expand the use of renewables and natural gas, and coordinate energy and environmental policy.

Energy efficiency

China has given a warm welcome to international cooperation in energy efficiency. Foreign donors and investors have financed new institutions, feasibility studies, project implementation, and joint ventures to improve energy efficiency through policy reform, boiler technology upgrades, district heating, and oil refinery retrofits.

The Beijing Energy Efficiency Center (BECon)

BECon is an example of an institution created to promote energy efficiency. BECon is a nonprofit, independent organization designed to achieve the twin goals of economic development and environmental protection. The center builds on a strong analytical base to promote energy efficiency in four areas: policy research and development, private business development, training and demonstration projects, and public education.

BECon was founded by Pacific Northwest Laboratory; World Wildlife Fund (WWF); Lawrence Berkeley Laboratory; and the Energy Research Institute (ERI) of the State Planning Commission of China. Like the energy efficiency centers established by Pacific Northwest Laboratory and WWF in Russia and Eastern Europe, BECon is directed and staffed by local energy experts. BECon's board of directors includes the founders and leading Chinese energy policy makers, managers, and researchers. The center is administratively affiliated with ERI, but its goal is to evolve into an independent institution as the Chinese political system changes. Initial funding is being provided by the U.S. Department of Energy, the U.S. Environmental Protection Agency, and WWF. BECon's goal is to become self-supporting after four years.

BECon is currently working on three projects to help improve energy efficiency in China. The first is helping to initiate a pilot project in integrated resources planning (IRP) in the city of Shenzhen, a Special Economic Zone across the border from Hong Kong. IRP is a process that puts energy conservation investments on an equal footing with investments in new power plant construction and gives utilities the responsibility for promoting energy conservation to meet at least part of their growing demand. The second project is coordinating an energy efficiency standards training project. BECon will work with Chinese policymakers and foreign experts to develop a curriculum on designing and implementing efficiency standards for electrical appliances. BECon's third project is working with Beijing city officials and transportation experts to develop a practical transportation plan that will combine mobility, energy efficiency, and environmental protection.

The Center also provides direct assistance to firms wishing to pursue energy efficiency joint ventures in China. BECon recently helped energy efficiency firms assess China's market potential by reporting on the opportunities for energy conservation in China. The resulting paper addresses technical issues related to boilers, fans, furnaces, and small cogeneration units, as well as suggesting policies to remove economic and systemic barriers to energy efficiency in China. BECon has also established connections with U.S. companies and discussed marketing energy efficiency technologies in China with them. These connections are a vital part of BECon's effort to promote private energy efficiency business ventures and investment in China.

United Nations and Multilateral Development Bank Efforts

Another effort to improve energy efficiency is the United Nations Development Programme and Global Environment Facility project on greenhouse gas emissions control.[63] The first part of the project is a feasibility study of improving the efficiency of 400,000 coal-fired industrial boilers in China by increasing the size and efficiency of the next generation of boilers. The second part is a case study in the Taiyuan area of Shanxi Province that will provide concrete investment recommendations for coal supply and use throughout China.

The Asian Development Bank has also shown an interest in assisting with energy conservation and environmental protection efforts in China. The Asian Development Bank is promoting demand side management in Jiangsu Province under a $479,500 technical assistance program financed from the Japan Special Fund for a Power Subsector Energy Conservation Study. The main objective is to review the organization of the power subsector in Jiangsu Province. Technical and financial aspects will be studied to identify barriers to successful introduction of demand side management electricity-saving measures.[64] The Asian Development Bank is also sponsoring a Harvard study that will develop environmental criteria for energy investments,[65] and has started to incorporate IRP into its energy planning process.[66]

The multilateral lenders have also put their money into improving industrial efficiency. The Asian Development Bank approved a $107 million loan for improving energy efficiency in Chinese fertilizer, cement, and steel industries in 1992,[67] and the United Nations Development Programme and the World Bank are funding projects to improve efficiency in the cement industry and the process equipment manufacturing sector.

Energy pollution control

A few smaller projects focus directly on environmental protection. The World Bank's Beijing Environment Project includes the Shijingshan Heating Supply project, which added 3.5 million square meters of district heating to the city, and the Environmental Protection Research Institute of Electric Power in Nanjing is finding low impact sites for coal-fired power plants by analyzing wind flows at potential sites with funding from the United Nations Development Programme and the World Health Organization.[68]

Renewable energy

China's soaring energy demand means that energy conservation alone will not reduce coal use. Only a combination of measures, including switching from coal to renewable energy sources, can achieve this goal. Renewable energy is an especially attractive alternative in remote rural areas located far from power grids.

The Solar Electric Light Fund (SELF), which promotes photovoltaics for rural people around the world, has begun working in the poor but sunny province of Gansu in western China. SELF is helping families living far from electric power grids to purchase solar electric systems that have three 8-watt DC lights, a power point for a television, and a 9 volt outlet for a radio cassette player.[69] Residents of the pilot village Majiacha reported a significant improvement in indoor air quality as a result of switching from kerosene to electric lamps. The environmental benefits of solar energy are even greater considering the avoided regional and global impacts of fossil fuel-fired electricity, which would be magnified by transmission losses incurred when transporting electricity from power plants to remote areas.

SELF's work was simplified because photovoltaic systems of international market quality were already available in China. Systems made by the Gansu Natural Energy Research Institute Photovoltaic Company (GGPV) were developed locally and use almost entirely Chinese-made components. GGPV agreed to be SELF's partner, supplier, and project contractor. Because the equipment was already provided, SELF's remaining tasks were then to convince people of the benefits of photovoltaic systems through demonstration projects and make the systems affordable by creating sustainable financing mechanisms.

SELF donated demonstration systems to two households in the village of Majiacha in 1992, and obtained $50,000 for a pilot project in Majiacha from the Rockefeller Foundation in January 1993. One hundred families (half of the village), convinced by the success of the demonstration systems, signed contracts to purchase systems for approximately $52 down and between two and five dollars a month until the systems were paid off. The seed money from Rockefeller thus created a revolving credit fund for future purchases. By November 1993, 100 systems had been installed, and SELF planned to install 12 more by the end of the year with money from the down payments held in the revolving fund.

SELF has set up a branch in Gansu Province called G-SELF, a Chinese foundation registered in Gansu Province and affiliated with SELF USA. The first phase of G-SELF's next project, "Seven County/1000-House National Exemplar Project for Solar Rural Electrification," began in September 1993 and is being funded by the W. Alton Jones Foundation. The United Nations Development Programme has recommended that expansion of G-SELF's activities be funded by the World Bank's Global Environment Facility. As in SELF's pilot project, sustainability is ensured through a revolving fund.

Payments are being collected by G-SELF's local subcontractor and project manager, Wang Wei Zhong, who is also the director of the Energy Company of the Rural Energy Office of Dingxi, the district in which Majiacha is located. The Chinese management decided that no interest should be charged in the pilot stage, and that the government should subsidize the project in its initial stages, lowering the cost of photovoltaic systems to consumers by 25 percent. The link to local government ensures that SELF gets paid for the systems—if counties fail to make their collections, the Dingxi district government can withhold credit funds for other projects. Mr. Wang has also initiated a program to help the 20 families who have made their down payment but are having trouble making their monthly payments. These 20 families have been loaned a pair of breeding pigeons, which will, according to project documents, produce $130 per year in baby pigeons that the Dingxi Rural Energy Office will collect and export, primarily to Hong Kong.

Other alternatives to coal

Natural gas and coal bed methane, like renewable energy, can help China reduce coal use and associated pollution. Unfortunately, most of China's natural gas has been used as inputs for the chemical fertilizer industry, leaving very little for coal substitution in the residential sector. Gas may become an important fuel for China, however, if current investment in the sector continues to grow and results in high yields. China's recent decision to open its onshore natural gas fields to foreign bidders may give the industry the push it needs.

Natural Gas

China has long been considered one of the world's largest unexploited regions for gas. Foreign companies were given access to China's offshore gas fields in the early 1980s, but China did not invite foreign participation in onshore fields until last year. The decision to expand foreign participation from offshore to onshore gas recovery was accompanied by a 50 percent increase in China's gas exploration and development budget and an increase its estimate of potential gas resources.[70] The Beijing Institute of Petrogeology now estimates potential reserves at 40 trillion cubic meters, not including 16 trillion cubic meters of potential coal bed methane reserves.[71]

Sichuan Province, with 360 billion cubic meters of proven reserves, now produces 43 percent of China's gas.[72] The World Bank recently approved a $250 million loan for exploration, development, and transmission of gas from existing fields in Sichuan Province. Bohai Bay also holds considerable proven reserves, approximately 230 billion cubic meters. China's first offshore gas field is in Bohai Bay.[73] It began operating late last year.

The best bets for large new finds are the South China Sea and Xinjiang Province in the far west. Arco. in collaboration with China National Offshore Oil Corp and a Kuwaiti partner, has already staked its claim to one field in the South China Sea with reserves of over 100 billion cubic meters.[74] That field is expected to go into operation by 1 January 1996. Tarim Basin reserves in Xinjiang Province are estimated at up to 2.5 trillion cubic meters, but transporting gas from this remote area is a major challenge.[75]

China claims that only 150 billion cubic meters of its gas reserves are commercially viable, but this estimate may be four or five times too low, according to at least one prominent U.S. expert. Robert Hefner of the GHK Company believes that China estimates are based on poor exploration and extraction techniques, and that up to 750 billion cubic meters of gas can be commercially recovered using modern technology.[76]

Coal bed methane

The U.S. EPA has been promoting the extraction of methane from coal mines in China, a process which provides clean fuel while reducing the threat of mine explosions and global warming. EPA has succeeded in obtaining funding for coal bed methane recovery projects from World Bank's Global Environment Facility.[77]

Policy assistance

Policy assistance is another important element of the foreign effort to reduce China's energy pollution. Foreign donors have promoted sustainable energy policy by directly aiding government projects, sponsoring associated activities, and funding interagency cooperation.

Integrated rural energy program

The Energy Sector Management Assistance Programme, a joint program of the World Bank and the United Nations Development Programme, assisted China's Ministry of Agriculture in implementing integrated rural energy policy. The goal was to develop technologies and methodologies for the Seventh Five Year Plan (1986-1990) key project "Rural Energy Comprehensive Trial in 12 Counties." The project has expanded in the Eighth Five Year Plan period (1991-1995) to become the key project "Integrated Rural Energy Construction in One Hundred Counties." Each county involved in the project develops an energy strategy that incorporates the specific environmental needs and potential of the area. Use of appropriate renewable energy, reforestation, and energy conservation is encouraged. The Rockefeller Foundation is supporting collaborative work by the University of Pennsylvania and the Ministry of Agriculture on monitoring, evaluation, and commercialization of key technologies developed under the project.

The One Hundred Counties project received funding from many sources. The state supplied $6100 annually in each county, and a corresponding amount was matched by the provincial and prefectural levels and each county provided twice as much as the state. [The combined funding totals approximately $ 14 million.] The counties use these funds mainly to formulate planning, technological training, pilot projects, demonstrations, technical service and information exchange. Builders and loans cover most of the implementation costs of projects.[78] The Bank of Agricultural Credit and the Industrial and Commercial Bank each provided $8.7 million in loans.[79]

Coordinating energy and environmental policy

The United Nations Environment Programme is promoting interagency cooperation by funding another effort to integrate energy and environmental policy. China's National Environmental Protection Agency is implementing the project with support from the Energy Research Institute of State Planning Commission and the Modern Policy Research Center for Economics and Environment. The purpose is to promote policies to reduce energy-related emissions of oxides of sulfur and nitrogen, particulate matter, and carbon dioxide, help develop sound energy strategies in specific regions, and promote energy efficiency and renewable energy. The plan is to develop national guidelines and concrete action plans for two to three regions that will include timetables and technology and funding requirements. The plans will be based on a quantitative analysis of environment impacts of energy use. Energy and environmental scientists and officials from several ministries and research institutes will use planning models to develop baseline and alternative scenarios that show the economic and environmental consequences of various policy options, such as promoting efficiency, fuel switching, coal washing, briquetting, gasification, and combined heat and power.

The project began in June 1993 and will be completed by November 1995. During the course of the 30 month process, the national coordinating committee, composed of representatives from Ministry of Electric Power Industry, Ministry of Finance, State Planning Commission, State Science and Technology Commission, Committee on Energy and Environment, and relevant universities, may use preliminary findings as the basis for policy suggestions.

A similar project was conducted by the municipal planning authorities of the northeastern city of Harbin from 1985-87. See Lu Yingzhong's report to the European Community, "Comprehensive Assessment of Impact of Energy on Environment in Harbin," for details.[80]

Conclusion

China's energy-related pollution problems cannot be easily overlooked. They are likely to plague China for years because of the country's great potential for growth in energy use. However, China's success in decoupling energy consumption and economic growth in the 1980s is cause for hope. Continued efforts to improve energy efficiency, use more renewable energy, and incorporate environmental factors into energy planning can help put China on a more sustainable energy path. Continuing and expanding international cooperation, such as the activities described in the case studies above, serve our common goals of reducing global carbon emissions, creating business opportunities, and improving the quality of life of Chinese citizens.

ENDNOTES

1. Yingzhong Lu, *Fueling One Billion* (Washington, DC: Washington Institute Press, 1993) 97.
2. Jonathan Sinton, ed., *China Energy Databook,* LBL-32822; UC-350 (Berkeley, CA: Lawrence Berkeley Laboratory, November 1992) II-5.
3. *Economic Daily (London, 26 March 1993).*
4. Sinton, II-5.
5. Ho Yi, "Problems of Oil Production," *Window* (7 August 1992).
6. J. Sathaye and N. Goldman, ed., *CO_2 Emissions from Developing Countries: Better Understanding the Role of Energy in the Long Term, Vol. III: China, India, Indonesia, and South Korea,* LBL-30060, UC-350 (Berkeley, CA: Lawrence Berkeley Laboratory, July 1991) 5.
7. "EPA-China Efficient Refrigerator Project," Project Document (Washington, DC: Environmental Protection Agency, 10 March 1993) 1 and *Energy in China* (Beijing: Ministry of Energy, 1992) 131.
8. Feng Liu, *Energy Use and Conservation of China's Residential and Commercial Sectors: Patterns, Problems, and Prospects* (DRAFT) (Berkeley, CA: Lawrence Berkeley Laboratory, August 1992).
9. Han Jiuling, "Air Pollution Control Policy and Measures in China (unpublished)," *U.S.-China Conference on Energy, Environment, and Market Mechanisms* (Washington, DC: Battelle, Pacific Northwest Laboratory, October 1992) 1.
10. Robert M. Wirtshafter and Chang Song-Ying, "Energy Conservation in Chinese Housing," *Energy Policy* (April 1987) 167.
11. *China: Reform and the Role of the Plan in the 1990s* (Washington, DC: World Bank, 1992).
12. "Growth of Nonstate Share of Industrial Output," *Associated Press-Dow Jones in China News Digest* (18 September 1993).
13. Robert M. Wirtshafter and Ed Shih, "Decentralization of China's Electricity Sector: Is Small Beautiful?" *World Development,* Vol. 18, No. 4 (1990) 505-512.
14. "Ten Top Issues In China's Environment," *China Environment News* (Beijing: September 1993).
15. Sinton, VII-2.
16. Sinton, VII-7.
17. Han, 1.
18. Han, 2.
19. Sinton, VII-11.
20. Sinton, VII-2.
21. J.L. Mumford, et al., "Lung Cancer and Indoor Air Pollution in Xuan Wei, China," *Science,* Vol. 235, (9 January 1987) 217-220.
22. Baruch Boxer, "Getting Beyond Rio," *Issues in Science and Technology,* Volume IX, Number 2, (1992) 46.

23. Xu Zhaoyi, et al., "The Health Effect of Air Pollution on Citizens in Liaoning Cities" (unpublished), *Conference for Environment and Climate and Change: The Challenge for China,* sponsored by World Wide Fund for Nature and the State Meteorological Administration of China (Beijing: 15 April 1991) 5.

24. All references to Chinese currency (yuan) in this paper have been replaced with an approximate U.S. dollar figure which may not be accurate due to widely fluctuating exchange rates. In all cases, yuan have been converted to dollars at the 1993 exchange rate of 5.75 yuan per dollar.

25. Sinton, VII-2.

26. Han, 1-2.

27. Shen Longhai and Zhou Changyi, "Overview of Economic Development, the Present Situation and Prospects for Environmental Protection and Energy Conservation in China," (unpublished) *U.S.-China Conference on Energy, Environment, and Market Mechanisms* (Washington, DC: Battelle, Pacific Northwest Laboratory, October 1992).

28. Liu Xueyi, personal communication, (Beijing: Energy Research Institute, State Planning Commission of China, 14 December 1993).

29. Ibid.

30. Norman L. Brown and Janet Welsh Brown, "Energy and Pollution in China," Proceedings from the *Conference on the Foreign Relations of China's Energy Policy,* Wingspread Conference Center, Racine, WI, 1-4 August 1991 (Washington, DC: American Enterprise Institute, 1992) 7.

31. Xiong Jiling et al., "Sources and Effects of Air Pollutants in Guizhou Province" (unpublished), *Conference for Environment and Climate and Change: The Challenge for China,* sponsored by World Wide Fund for Nature and the State Meteorological Administration of China (Beijing: 15 April 1991) 2.

32. Ibid.

33. "Guangxi's Environment Worsens," *China Environment News* (September 1993).

34. William U. Chandler, Alexei A. Makarov, and Zhou Dadi, "Energy for the Soviet Union, Eastern Europe and China," *Scienttfic American* (September 1990) 125.

35. Mike Hulme, et al., *Climate Change due to the Greenhouse Effect and its Implications for China,* (Gland, Switzerland: WWF-World Wide Fund for Nature (formerly World Wildlife Fund), 1992).

36. Ibid.

37. Wang Futang, et al., "A Modelling Experiment of Effects of Climate Change on the Food Production in the Major Agricultural Areas of Eastern China" (unpublished), *Conference for Environment and Climate and Change: The Challenge for China,* sponsored by World Wide Fund for Nature and the State Meteorological Administration of China (Beijing: 15 April 1991).

38. Liu Zhenxia, *Modern Sea Level Changes of the Eastern China Seas and Their Influences on Coastal Areas* (Qingdao, China: First Institute of Oceanography, 1991).

39. Hulme.

40. Jessica Hamburger, *U.S.-China Conference on Energy, Environment and Market Mechanisms: Summary Papers,* Global Studies Program Report (Washington, DC: Battelle, Pacific Northwest Laboratory, July 1993) 11-12.

41. Alan J. Krupnick, "Urban Air Pollution in Developing Countries: Problems and Policies," (Washington, DC: Resources for the Future, June 1991).

42. Hamburger, 11-12.

43. Ibid.

44. Ibid.

45. "Coal Mining in North China Plain Cripples Environment," *United Press International in China News Digest* (13 October 1993).

46. Hamburger.

47. "The Coal Sector in China," *Asian Energy News* (Bangkok: Asian Institute of Technology, May 1993).

48. "New Austerity Program a Critical Test for China's Government," *Asian Digest* (South Orange, NJ: International Trade and Exchange Group, August 1993) 2.

49. David Schneider, et al., "Power Plays," *The China Business Review* (November-December 1993).

50. Ibid.

51. Xin Dingguo, personal communication (Beijing: Energy Research Institute, 16 December 1993).

52. Ibid.

53. Ibid.

54. Dai Yande, personal communication (Beijing: Energy Research Institute, 20 December 1994).

55. Ibid.
56. Lester Ross, *Environmental Policy in China,* (Bloomington: Indiana University Press, 1988) 131-175.
57. Abigail R. Jahiel, "The Deng Reforms and Local Environmental Protection: Implementation of the Discharge Fee System, 1979-1991" (unpublished), *46th Annual Meeting of the Association for Asian Studies,* Boston, 24-27 March 1994.
58. "Air Pollution Prevention and Control (Daqi wuran fingzhi gongzuo)" (in Chinese), *China Environmental Yearbook,* 1990 (Zhongguo huanjing nianjian), (Beijing: National Environmental Protection Agency, 1991).
59. Liu Xueyi, et al., *China's Sulfur Dioxide Emission Control Policies and Recommendations for Cooperation in Sulfur Elimination* (Zhongguo kongzhi SO_2 paifang de zhengce ji tuoliu hezuo de shexiang) (in Chinese), (Beijing: Energy Research Institute, March 1993).
60. Hamburger.
61. Geoffrey Crothall, "Beijing is Easier to Live In as Environment is Cleaner," *South China Morning Post* (20 December 1993).
62. Liu Xueyi.
63. "Issues and Options in Greenhouse Gas Emissions Control," Project Document, Project of the Government of the PRC, CPR/91/G32/A/IG/42 (Washington, DC: World Bank-Global Environment Facility, 1991).
64. ADB News Release No. 107/93, 25 June 1993, *Asian Energy News* (Bangkok: Asian Institute of Technology, July 1993).
65. Fiona Murray, personal communication (Cambridge, MA: Harvard University, January 1994).
66. Nicholas Lenssen, "All the Coal in China," *WorldWatch* (Washington, DC: World Watch Institute, March-April 1993).
67. Ibid.
68. Sid Kane, "China's New Environmental Ethic," *Sustainable Development and the Environment* (United Nations Development Programme).
69. "The MaGiacha Village PV Pilot Project, Gansu Province, China," (Washington, DC: Solar Electric Light Fund, November 1993).
70. "Plenty in Reserve," *China Economic Review* (February 1994) 18.
71. Ibid.
72. Ibid.
73. "First Offshore Gas Field Now Operating," *Process in Asian Energy News* (Bangkok: Asian Institute of Technology, October 1993).
74. "China Says Offshore Gas Field Should Open in 1996," *Journal of Commerce* (17 February 1994).
75. "China Reports Strong Foreign Interest in Tarim Oil and Gas," *Reuters in China News Digest* (17 August 1993).
76. Robert A. Hefner, letter to Rocky Mountain Institute (Oklahoma City: The GHK Company, 29 October 1992).
77. Hamburger.
78. "Rural Energy Development in China," (Beijing: Ministry of Agriculture, Department of Energy and Environmental Protection, 1993).
79. Liu Hongpeng, personal communication (Beijing: Department of Energy and Environmental Protection, Ministry of Agriculture, 4 October 1993).
80. Lu.

STATEMENT OF KENNETH L. LAY, CHAIRMAN AND CHIEF EXECUTIVE OFFICER, ENRON CORP.

Mr. Chairman, I am pleased to have this opportunity to address the Senate Energy and Natural Resources Committee on project development in the Pacific Rim nations. This region has achieved major economic growth in the last decade which in the years ahead shall require significant infrastructure development as its citizens increasingly achieve a higher standard of living.

I know that you and the members of this Committee have traveled to the region many times yourselves and realize that the leadership of these nations are racing to fulfill the full potential of their people. These nations have dramatically emerged on the world economic stage and are calling on United States industry to help them meet their economic challenges.

We as a nation through coordination between government and industry must use our best American skills in project development, technology and understanding of other cultures if we are to take advantage of this opportunity.

Our country has shown great initiative in facing the political challenges to achieve our national economic goal of greater export of U.S. sourced goods and services. Our initiatives in promoting free trade and open markets has benefited us and others. The recently concluded Uruguay Round, which, when ratified, will establish the World Trading Organization as a successor to GATT, promises to liberalize trade and open markets among the signatory nations. And the Asian Pacific Economic Council (APEC), thanks to the prompting of the United States, has recently been rejuvenated as a mechanism to address long-term trade and investment needs of the region. Additional tools the U.S. has successfully used to strengthen economic and political ties are Most Favored Nation (MFN) status and the General System of Preferences (GSP).

However, there are serious threats on the horizon. We are currently in a confrontation with China and Japan over differences which, if unresolved, could undermine the position of influence the United States now enjoys. It requires our greatest skill and wisdom to solve these problems with these two countries, which are important to us and others in the region who desire a liberalized trading regime.

Last November, I had the opportunity to address the China-U.S. Energy Conference of the Chinese-American Institute of Engineers and Scientists in Beijing.

I found an impressive group and one that is totally focused on finding the best technical and scientific solutions to developing its nation. The participants were enthusiastic and ready to make the necessary changes to achieve success for their nation, themselves and their fellow citizens.

Following the sessions, I had several days of meetings with many of the leaders of the Peoples' Republic of China who acknowledge their responsibilities, recognize their country's needs and are eager to have our participation in their growth.

China's dynamic economy grew 12.8 percent in 1992, according to Bank of America's country reports, and its momentum continued in 1993 due to foreign investment and consumer spending. The nations of Southeast Asia, particularly Indonesia, Thailand, and Vietnam, are likewise experiencing rapid economic growth. Foreign investment is being attracted to these countries by reason of favorable labor and land costs that help exports, trends in liberalizing the economy and by the large and expanding potential of the domestic consumer market.

These countries' economic progress is being built on a foundation of large, basic industries replacing a largely agrarian work force. There is a growing dependence on energy use, which is putting more and more citizens to work in an urbanizing economy with better paying jobs, skills and knowledge.

The World Bank (February 1990) outlook estimates that between now and the year 2000 China will require a total of 104 GW of new power plants, or about 18.4 percent of the projected new power generation needed around the world. More recent estimates run as high as 150 GW or about 20 GW per year from 1994 to 2000. The ASEAN countries expect to almost double power capacity, an increase of 44 GW by 2000. The breakdown by fuel source in China is (according to the World Bank):

	New GW installed
Hydro	35.6
Nuclear	3.9
Coal	64.3
Natural Gas	0.0
Total (gigawatts)	103.8

China is still at the threshold of making natural gas a strategically positioned component of its domestic fuels portfolio for the economic and social benefit of its citizens. The government's energy plans are to double its 1.5 BCFD of gas production in China by the year 2000. Today natural gas fuels just 2 percent of China's economy. Coal fuels 75 percent of its energy requirements and oil, the most rapidly growing fuel used in the country, today provides 2.3 MMbd of its energy—versus 1.8 MMbd of oil use in 1980, up 25 percent in this decade alone. At the current rate, China will become a net importer of oil and oil products by 1995.

China has about 50 Tcf of natural gas reserves—or about one-third as much as the U.S. proved gas reserves of 158 Tcf. Yet China today has only very limited natural gas pipelines—a total of about 3,900 miles (or 6,200 kilometers) compared to the U.S. with 280,000 miles (or 448,000 kilometers), which is 70 times larger. China will require a significant investment in the natural gas infrastructure to capitalize on these native resources.

Plentiful, accessible and cost-effective energy based on best available technologies can help the emerging economies of the Pacific Rim solve their domestic energy requirements. But along with the growth in energy consumption comes the need for a greater environmental concern. The U.S. also possesses the technologies required to ensure that economic growth does not come at the expense of the environment.

We at Enron believe that natural gas is the most environmentally benign fuel available. Tonnages of pollutants that can be saved by gas use for power plants in China versus high sulfur coal or high sulfur residual oil are enormous. If China burns 1.5 Bcfd of gas in 8 gigawatts of the 64 gigawatts of planned coal plants needed to be built (according to the World Bank) by the year 2000, it cuts its 37 million tons of 502 pollution annually from all power plants by an estimated 4.5 million tons or by approximately one-eighth.

When compared to a new 240 MW coal-fired power plant that meets U.S. EPA emissions standards, a new combined-cycle plant

- reduces SO_2 pollution by 100 percent
- reduces particulates by 96 percent
- reduces CO_2 by 55 percent
- reduces NO_X by 85 percent
- eliminates 168,000 tons of sludge and 60,000 tons of ash annually.

The economic rationale for use of natural gas is compelling. The capital cost of a natural gas combined-cycle power plant is less than $700 per KW installed versus $1,820 per KW for a coal plant. Also, gas combined-cycle plants have 25 percent greater thermal efficiency than coal plants. With the advantage of lower capital cost, a great portion of these savings can be spent to build a gas grid to support the development of gas reserves around China. Also, pipeline field operations needed will create new permanent technical jobs in remote areas.

Favorable gas combined-cycle power plant economics can integrate pipeline grid economics (and clean up air pollution) instead of the use of coal plants with stack gas scrubbers.

At Enron we feel strongly that the advantages of generating power from gas fired combined-cycle plants are overwhelming. They are cleaner, cheaper and operate with greater availability for dispatch than either the coal or nuclear options. When considering that a growing large economy like China can obtain the extra benefit of a gas pipeline grid for the same "price tag" as a modern coal or nuclear plant alone, it is a realization that we believe shall soon become a reality for the Chinese energy economy.

Natural gas supplies brought to market in China's diverse regional locations would also save long haul transport costs for coal and create jobs in a new and growing gas industry to help keep local regional economies strong. This applies not only to traditional sources of supply but also to non-traditional sources of natural gas such as coal seam gas.

Enron is a leader internationally in the development of coal seam gas. A recently issued study by the National Petroleum Council (an industry advisory body appointed by the U.S. Secretary of Energy) estimates that fully 40 percent or 520 Tcf of the remaining economically recoverable natural gas resources in our country are expected to come from the non-traditional sources. We believe similar potential exists in China.

We believe that the U.S. gas industry has the technological and competitive edge and that we are well positioned to market our expertise abroad in all parts of the world, and most particularly China. Enron is a technology leader in providing services from the gas wellhead to the electric power busbar and products which help remedy problems of oil and coal use and air pollution control through natural gas development. And, we market our services worldwide.

At Enron Operations Corp., we've done large $1 billion projects like the 1.7 Bcfd Northern Border pipeline, which carries Canadian gas under cold, icy weather conditions; and we built the connector pipeline to the 1.4 Bcfd U.K. CATS pipeline in North Sea waters.

Last year, an Enron-led consortium acquired the largest gas pipeline system in South America, Argentina's Transportadora de Gas del Sur gas pipeline system. Since the acquisition, Enron has applied to the Argentina pipeline technical automation skills which have evolved during our 50+ years of pipeline operation in North America. A gas control center has been installed and operates the Argentina pipeline using a computer system based on Enron's U.S. pipeline system operations and software. In addition, Enron is already in the process of expanding the pipeline capacity by 16 percent to be ready for Argentina's next cold weather season.

Enron also completed the largest natural gas cogeneration plant in the world in the United Kingdom. That 1,875 megawatt plant is currently capable of providing more than 4 percent of all the electricity used in the United Kingdom. Enron's involvement in Argentina and the United Kingdom is typical of the activities we pursue worldwide and is a good example of the world leadership the U.S. energy industry maintains and the scale and type of clean energy technologies we are attempting to transfer to China and other Pacific Rim countries.

In China and the Pacific Rim, Enron has already built a sizable, growing and important business:

Enron Oil and Gas is involved in exploration in three different areas of China using coalbed methane recovery technology developed through the significant advances made in the U.S. over the last five years.

On the conventional side, Enron has been the leader in developing existing oil and gas fields using a low cost/high tech approach, similar to what we are applying today offshore in Trinidad and will soon be applying offshore India. This kind of approach can be applied both offshore and onshore, and we are actively pursuing these kinds of opportunities in China.

Enron is marketing clean LPGs in all the large Asian markets, especially in the very rapidly growing coastal and southern regions of China.

Methyl tertiary butyl ether (or MTBE) blending with gasoline reduces pollutants. Through our 15,000 b/d world-scale size MTBE plant near Houston, Texas, Enron manufactures MTBE feedstock for blending into reformulated gasoline to clean up the air pollution in major urban areas. Natural gas is, of course, the feedstock for methanol, the main component of MTBE. At 130 octane value with cleaner tailpipe emissions, MTBE is an ideal product aimed squarely at remedying the dual problems of oil imports and air pollution. Enron also markets MTBE in Asia, and has targeted the gasoline market in China as a strategic growth market.

Enron has been very active in the Pacific Rim region. In the last year, Enron has completed two power plants in the Philippines. And Enron is currently developing several integrated gas infrastructure projects, from gas production to pipelines to power plants in China, Indonesia, and Thailand. We are also investigating similar opportunities in Australia, New Zealand, Korea, Taiwan and now even Vietnam.

At Enron, we believe we are entering a new era—the new world of natural gas. We live in a world poised for economic growth but aware of our Earth's fragile nature and thus seeking clean-burning fuels for a better environment. China is key to this effort.

Enron Corp. is at the forefront of this new world. We offer reliable, competitively priced and abundant natural gas fuels packaged to meet the individual needs of customers on almost every continent. Asia, and particularly China, represent an expansive market, perhaps, the largest for the technologies and services we represent.

As one of the world's largest integrated natural gas companies, Enron Corp. is positioning itself to meet growing energy needs throughout the world. Our company operates natural gas pipelines on two continents and handles about 20 percent of the natural gas consumed in the U.S. Enron also is the largest buyer and seller of natural gas in North America and is a leading producer and marketer of natural gas liquids and MTBE worldwide. Through its power segment, Enron is one of the largest independent developers and producers of electricity worldwide with facilities in Europe, the Americas and the Pacific Rim, with ownership in approximately 3.7 gigawatts of operating plants. Through its 80 percent ownership in Enron Oil & Gas Company, Enron is a leading independent producer of natural gas in the U.S., Canada, and Trinidad; and has additional development projects throughout the world using its coalbed methane technology group and offshore and onshore gas development expertise.

Enron Corp. is well on its way to achieving its vision of becoming the world's first natural gas major—the most innovative and reliable provider of clean energy worldwide for a better environment.

We are looking forward to applying our capabilities to meeting the rapidly growing energy needs of China and the other Pacific Rim developing nations during the remainder of this century and into the next century. These markets provide enormous opportunities for the U.S. natural gas industry and to Enron.

Mr. Chairman and members of the Senate Energy and Natural Resources Committee—

This Committee which has heard testimony about the needs of the Pacific Rim nations in infrastructure along with its own firsthand knowledge of the region must convey its findings to the highest level of our government or we could face long-term disruption of our national interest in Asia.

Near term, we as a nation should extend Most Favored Nation (MFN) status to the People's Republic of China and work to resolve our differences amicably with Japan.

The Economist of March 12, 1994 reported that the U.S. exports $9 billion worth of goods to China which accounts for at least 200,000 American jobs.

The United States has gained and continues to gain in convincing nations that open trade helps bring mutual benefits. These benefits are both economic and political.

I congratulate you for this hearing which has focused on the value of development in the Pacific Rim. It's important to our long-term well being and those of other nations.

Thank you for allowing me to express my views.

○